DEVELOPING SCIENCE LANGUAGE

for

Living Things

with

8 – 9

year olds

Neville Evans

Published by Scholastic Ltd,
Villiers House,
Clarendon Avenue,
Leamington Spa,
Warwickshire CV32 5PR
Visit our website at www. scholastic.co.uk

Printed by Alden Group Ltd, Oxford

1 2 3 4 5 6 7 8 9 0   2 3 4 5 6 7 8 9 0 1

AUTHOR
Neville Evans

LITERACY CONSULTANT
Gill Matthews

EDITOR
Joel Lane

ASSISTANT EDITOR
David Sandford

DESIGNER
Erik Ivens

SERIES DESIGNER
Rachael Hammond

COVER PHOTOGRAPH
© Stockbyte

ILLUSTRATIONS
Ann Kronheimer

The author would like to acknowledge the help
of Kath Phelan as Science Consultant.

British Library Cataloguing-in-Publication Data
A catalogue record for this book is available from the British Library.

ISBN 0-439-01877-3

Designed using Adobe Pagemaker

CONTENTS

5 Introduction
6 Word list

7 Teeth ...... (description/explanatio

8 Teeth questions...... (higher level)
9 Teeth questions...... (lower level)
10 What do teeth do? ...... (matching)
11 A set of teeth...... (identifying, writing, drawing)
12 Whose teeth are these?...... (matching, describing)
13 Find the words...... (wordsearch, explaining)
14 Fill the gaps...... (writing, drawing)

15 Diets ...... (description)

16 Diets questions...... (higher level)
17 Diets questions...... (lower level)
18 Find the foods...... (identifying, planning)
19 Hidden words...... (wordsearch, writing)
20 Types of food...... (grouping)
21 How many units?...... (using data, calculating)
22 What do animals eat?...... (researching, writing)

23 Health care ...... (description, explanation)

24 Health care questions...... (higher level)
25 Health care questions...... (lower level)
26 Feel the pulse 1...... (generating questions)
27 Feel the pulse 2...... (planning an investigation)
28 Body and brain exercises...... (interpreting, writing)

29 Plant systems ...... (description)

30 Plant systems questions...... (higher level)
31 Plant systems questions...... (lower level)
32 Which part does what job?...... (matching, writing)
33 The hidden plant...... (using clues, drawing, writing)
34 Plant parts that we eat...... (completing sentences, grouping, writing)
35 Look very closely...... (observing, generating questions)
36 A plant dictionary...... (following instructions)

37 Plants and their needs ...... (description)

38 Plants and their needs questions...... (higher level)
39 Plants and their needs questions...... (lower level)
40 True or false?...... (answering and generating questions)
41 Healthy and unhealthy...... (describing, writing)
42 Gardener's world...... (choosing words, writing instructions)
43 Dr Gardener...... (problem solving, explaining)
44 Good for plants and good for humans?...... (wordsearch, comparing)

45 The human skeleton ...... (description)

46 The human skeleton questions...... (higher level)
47 The human skeleton questions...... (lower level)
48 Buried bones...... (wordsearch, describing)
49 Skeleton key...... (labelling, describing)
50 Word tower...... (solving puzzle, writing acrostic poem)
51 Skeleton card game...... (matching sentence starters and endings)
52 Make a skeleton puppet 1...... (writing instructions)
53 Make a skeleton puppet 2...... (constructing, describing)
54 How our bones protect us...... (labelling, writing)

CONTENTS

55 Other animal skeletons ...... *(description)*

56 Other animal skeletons questions...... *(higher level)*
57 Other animal skeletons questions...... *(lower level)*
58 Like and unlike...... *(listing similarities and differences)*
59 Whose skeleton is it?...... *(matching words and pictures)*
60 Odd one out...... *(selecting, explaining)*

61 Human muscles ...... *(description, explanation)*

62 Human muscles questions...... *(higher level)*
63 Human muscles questions...... *(lower level)*
64 Missing muscle words...... *(completing sentences, solving clues)*
65 Bend and stretch...... *(labelling, writing)*
66 Muscular activities...... *(writing labels)*

67 Habitats ...... *(description)*

68 Habitats questions...... *(higher level)*
69 Habitats questions...... *(lower level)*
70 Changing places...... *(matching, sketching, writing)*
71 What lives where?...... *(completing sentences, writing)*
72 crambled words...... *(reordering letters, explaining)*
73 True or false?...... *(answering questions, rewriting)*
74 Habitat puzzle...... *(answering clues, writing factfile)*

75 Keys ...... *(description)*

76 Keys questions...... *(higher level)*
77 Keys questions...... *(lower level)*
78 What shape...... *(using a key)*
79 What is it?...... *(identifying with a key)*
80 Making a key...... *(writing, explaining)*
81 A key for plants...... *(identifying plants)*
82 A key for birds...... *(identifying plants)*

83 Food chains ...... *(description, explanation)*

84 Food chains questions...... *(higher level)*
85 Food chains questions...... *(lower level)*
86 Pond life...... *(matching words, describing)*
87 Producers and consumers...... *(sorting, explaining)*
88 Arranging a food chain...... *(identifying, writing)*
89 Chains and webs...... *(writing labels)*
90 Food chains crossword...... *(solving clues, writing flow charts)*

91 Conservation ...... *(description, explanation)*

92 Conservation questions...... *(higher level)*
93 Conservation questions...... *(lower level)*
94 Careless humans...... *(describing, writing a letter)*
95 When do birds need food?...... *(handling data, making a chart)*
96 A nest box for a robin...... *(writing instructions, explaining)*

INTRODUCTION

Children often struggle to remember science words. Sometimes the words seem strange or unusual, and sometimes the words we use in science have other meanings in everyday life. Think about these science words: *sense, animals, exercises.* If you ask a child what these words mean, you are likely to get responses such as: 'People who talk silly do not make sense'; 'Animals are things with four legs, like cats and dogs'; 'I do my exercises in my green book'. But when children go into science lessons, we sometimes assume that they already understand that 'sense' refers to our five faculties; 'animals' includes birds, fish, insects and humans; and 'exercises' are physical routines to keep our bodies healthy.

Scientific language

This series aims to give children practice in using science words, both through science activities and in 'real life' contexts, so that they become familiar with the meanings of these words. Use of correct scientific vocabulary is essential for high attainment in national assessment tests. The QCA *Scheme of Work for Science* for Key Stages 1 and 2 in England suggests vocabulary for each of its units; although these books are not divided into exactly the same topics, the QCA vocabulary and its progressive introduction are used as the basis for the word selection here.

The science covered is divided into units based on topics from the national curricula for England, Wales, Scotland and Northern Ireland. In this book, the science is drawn from the 'Life processes and living things' statements for ages 8–9 relating to life processes, humans and other animals, green plants, variation and classification and living things in the environment. The boxed letters at the bottom of each page show to which curriculum the focus of each activity relates. For example, for the activity on page 59, the boxes E NI W S indicate that the activity focuses on a topic from the Scottish Guidelines only.

Science and literacy

The National Literacy Strategy for England suggests teaching objectives and gives examples of the types of activities that children should encounter during each year of primary school. This book uses many of these techniques for developing children's understanding and use of scientific language. The activities here are mainly intended for use in science time, as they have been written with science learning objectives in mind. However, some of the activities could be used in literacy time. Science texts have already been published for use in literacy time, but many of them use science content appropriate for older children.

During literacy time you need to be focusing on language skills, not teaching new science. It is with this in mind that these sheets, drawing from age-appropriate science work, have been produced. It is also suggested that these sheets are used in literacy time only after the science content has been introduced in science time.

The series focuses mainly on paper-based activities to develop scientific language, but it is hoped that teachers might also use some of the ideas in planning practical science activities.

About this book

Each unit in this book begins with a non-fiction text that introduces some key scientific vocabulary. The key words are highlighted by bold type. The texts cover a range of non-fiction genres.

Following this text are two comprehension activities that help children to identify and understand the key words (and introduce some new science words). They are pitched at two levels:

 for older or more able children

 for younger or less able children.

Although the comprehension activities are designed to be used mainly during science time, you may wish to use the texts as examples of non-fiction texts in literacy time. The comprehension pages contain two or three types of question (a change of icon indicates a change in the type of question):

 The answer can be found in the text.

 Children will need to think about the answer. These questions usually elicit science understanding beyond what the text provides.

 An activity aimed at developing the children's literacy skills. These are optional extension activities for individual or group work, with teacher support if necessary.

Following the comprehension pages in each unit are activities aimed at developing children's understanding and use of the key vocabulary and additional science vocabulary. Strategies used include: matching pictures and labels, wordsearches, explaining terms, making labelled diagrams, planning, grouping, handling data, generating questions, solving clues, writing instructions, writing a poem, matching sentence starters and endings, writing comparisons.

WORD LIST

active	feed	meat	space
adapt	finger	milk teeth	species
adapted	fish	minerals	spine
adequate	flavour	molar	starch
air	flower	move	stalk
animal	food chain	muscle	stem
	food web		stickleback
backbone	fracture	natural	sugar
balanced	fruit	nutrition	sunlight
biceps	function	information	survive
bite			symmetrical
brain	greenhouse	omnivore	
	grow	organism	tear
calories	gums	organ	tendon
canine		oxygen	tissue
carnivores	habitat		tone up
chew	healthy	pattern	tooth decay
climate	heart	permanent	triceps
coloured	height	petal	
conditions	herbivore	plant	units
conservation		plant systems	unvaried
conserve	identify	predator	
consumer	inadequate	premolar	varied
contract	incisor	prey (on)	variety
cycle of life	infection	producer	vegetable
	injury	proportion	vertebrate
detail	internal	pulse	visible
diet	invertebrate	ribs	
difference	invisible	root	warmth
different			waste
	jaw	scented	water
elbow	joint	sense	woodlice
energy		shape	wrist
exchange	key	shell	
exercise	kilojoule	shoulder	
external	knee	similar	
extinct		similarity	
	leaf	skeleton	
fat	life cycle	skin	
feature	lungs	skull	

Teeth

An adult has 32 **permanent** teeth in a complete set, 16 in each **jaw**. At the front of the mouth are the **incisors**. These have sharp edges to bite off pieces of food. On both sides of the incisors are the **canines**. These are strong and pointed to tear food. At the back of the mouth are the **premolars** and the **molars**. These are used to chew and crush food.

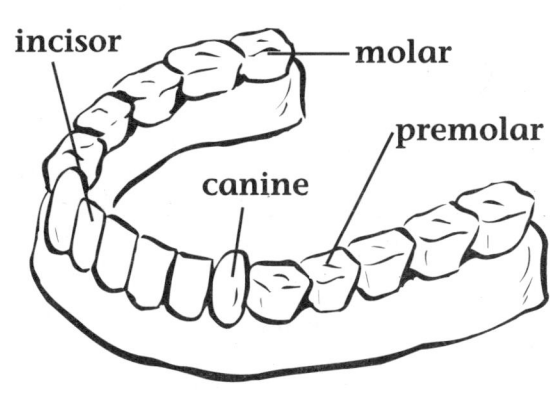

incisor — molar

premolar

canine

Humans have two sets of teeth during their lives. The baby teeth or **milk teeth** appear between the ages of four months and two years. There are no molars at this stage, because very young children eat only soft foods. Between the ages of five and ten years the milk teeth fall out, usually one or two at a time. They are replaced by the permanent teeth. This also happens to cats and dogs.

Very few adults have perfect teeth and **gums**. Most **tooth decay** is caused by eating sugary foods. We need to clean our teeth regularly to prevent tooth decay and gum **infection**. It is very important for us to brush after meals in order to remove bits of food from the gaps between our teeth. We should also visit the dentist regularly to have our teeth and gums checked.

Teeth

1.   What are **canine** teeth used for? _____

2.   Look in a dictionary for the meaning of the word **canine**.

What animal does it mention? _____

3.   Describe two differences between our **milk teeth** and our **permanent teeth**.

a) _____

b) _____

4.   Which teeth have sharp edges? _____

5.   Name three foods that this kind of teeth deal with best.

_____ _____ _____

6.   How old are humans
when their milk teeth appear? _____

7.   What causes most **tooth decay**? _____

Describe two ways to prevent **tooth decay** and **gum infection**.

a) _____

b) _____

8.   If you eat lots of sweets and chocolate, your teeth will become unhealthy. Write a sentence to explain why.

_____

_____

Ask your friends two questions:
1. How many times each day do you brush your teeth?
2. When do you brush your teeth?
Make a table to show what they say.

Teeth

1. Cross out the incorrect word in each sentence.

Canine teeth are used to **chew/tear**.
Molars are used to **bite/chew**.

2. Write the missing word or number in each sentence.

There are no _____ in a set of **milk teeth**.

An adult has _____ permanent teeth.

3. Which teeth have sharp edges? _____

4. What does this kind of teeth do best? Choose one of these words.

        bite   tear   chew

5. How old are humans when their milk teeth appear?

6. When should we brush our teeth? _____

What person should we visit to have our teeth checked? _____

 7. Sweets are not good for our teeth. Write a sentence to explain why.

_____

_____

Ask your friends two questions:
1. How many times each day do you brush your teeth?
2. When do you brush your teeth?
Make a table to show what they say.

Name	How many times	When?

What do teeth do?

Draw a line to match each type of tooth to its job.

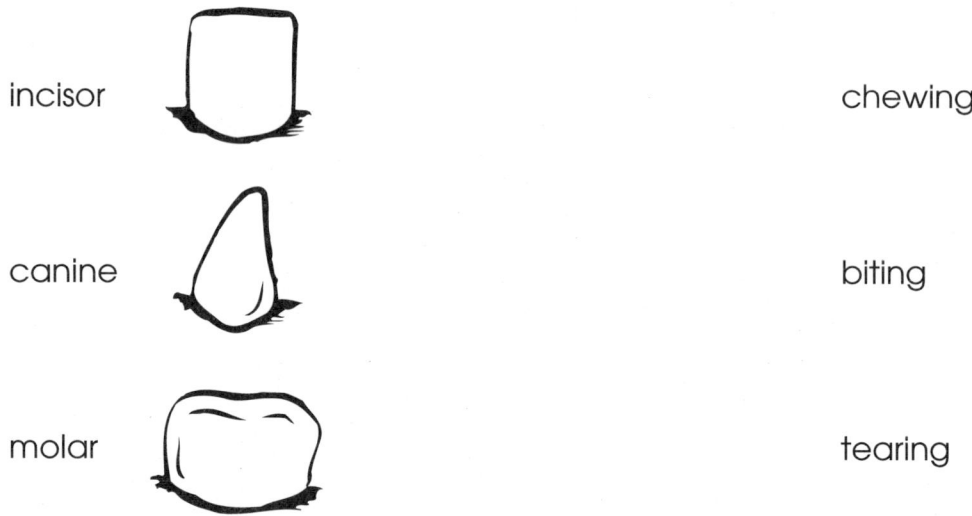

incisor   chewing

canine   biting

molar   tearing

Which teeth are the main ones used for each food?
Use these words for your answers:

incisor   canine   molar   none

Food	Which teeth?
apple	
bread	
milk	
biscuit	
meat	

Think of two more foods. Fill in the spaces in the table.

A set of teeth

Daniel has lost three of his milk teeth.

A

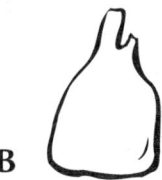

B

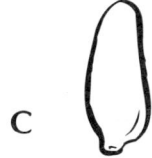

C

Show where each of Daniel's lost teeth came from by writing **A**, **B** and **C** in the correct gaps.

For each lost tooth, write what kind of tooth it is and what it is used for.

Tooth A is _____. It is used for _____.

Tooth B is _____. It is used for _____.

Tooth C is _____. It is used for _____.

On another sheet of paper, make a large diagram of your friend's teeth. Label the different kinds of teeth.

If there are any missing teeth, label them A, B, C and so on. Then write down which kind of tooth each missing tooth was.

Whose teeth are these?

Draw a line to connect each animal to the description of its teeth.

dog

sheep

rabbit

human baby

shark

It has lots of sharp teeth to tear meat.

It has no canines, but it has a hard pad in the roof of its mouth to help it chew grass.

It has very large canines and strong molars.

The incisors grow throughout its life, but it does not have any canines.

It has no molars.

Choose two other animals and write descriptions of their teeth.

1. _____

_____

_____

2. _____

_____

_____

SCHOLASTIC DEVELOPING SCIENCE LANGUAGE for *Living Things* with 8–9 year olds

Find the words

d	e	n	t	i	s	t	i
m	o	l	a	r	g	s	n
h	t	e	e	t	h	m	c
c	b	i	t	e	e	u	i
e	w	e	h	c	r	g	s
f	b	r	u	s	h	x	o
n	d	e	c	a	y	y	r
i	z	e	n	i	n	a	c

Draw lines in this wordsearch to show ten words. The words are all about our teeth. The words may be horizontal or vertical. Some may be upside down or back to front.

Write the words below. A picture clue and the first letter have been provided for each word.

1. g_____

2. b_____

3. d_____

4. b_____

5. i_____

6. m_____

7. t_____

8. c_____

9. d_____

10. c_____

Choose three of the words. Explain what they mean.

_____

_____

_____

Fill the gaps

Fill in this table by writing a word or a number, or drawing a picture, in each of the spaces.

Drawing of tooth	What type of tooth is this?	How many of these does a toddler have?	How many of these does an adult have?
	molar		

	incisor		

Write a sentence to answer each question.

1. Why do very small children not need molar teeth? _____

_____

2. Why does a rabbit need strong incisors? _____

_____

3. Why does a cat need strong canines? _____

_____

4. Why does a cow need strong molars? _____

_____

Diets

All **animals** need to **feed** so that they can **grow**, be **healthy** and be **active**. Different animals need to eat different types of food.

We use the word **diet** to describe the foods that we usually eat. Our **diet** needs to be **varied**, **balanced** and **adequate**.
- A **varied** diet means eating several different types of foods. One type of food alone will not keep us healthy.
- A **balanced diet** means eating the right amounts of different foods, not too much of one type and too little of another.
- An **adequate** diet means that we eat just enough food, not too much or too little.

  Different groups of foods help us in different ways:
- **Meat and fish** help us to grow and be strong.
- **Sugars and starches** help us to be active.
- **Fats** help us to keep warm and healthy.
- **Vegetables** help us to grow and keep healthy.
- **Fruits** help us to recover from illness and injury.

Different humans prefer to eat different kinds of foods – such as fish or pasta or cheese. However, all humans need to eat a variety of different foods. This is because humans are **adapted** to eating a varied diet. Some other animals are adapted to eating only one kind of food. **Carnivores** such as cats only eat meat or fish. **Herbivores** such as cows only eat plants. Bears, like humans, are **omnivores**: they can eat meat as well as fruits and vegetables.

Diets

1. What must all animals do to be healthy? _____

2. What is the science word
for the food we usually eat? _____

3. What are the three important words that describe a good diet?

_____   _____   _____

4. Write a sentence to explain what one of the three words means.

_____

_____

5. If a person's weekly diet is made up of apples and nothing else, what is wrong with it? Think of three important words.

_____   _____   _____

6. Which types of food help us to grow? _____

Which types of food help us to be active? _____

7. Does a cat have a varied diet? _____

Explain your answer. _____

_____

_____

8. What two things might happen to a person whose diet is not **adequate**?

_____

Make a chart to show the five main groups of foods. Write the reason why each food group is good for us. For each group, name two foods that we can buy from shops.

Diets

1. Complete these sentences.

To be healthy, all animals need to _____

The science word for the food we eat is our _____

2. There are three important words that describe a healthy diet. One of them is written below. Write the other two.

varied _____   _____   _____

3. What does an **adequate** diet mean? Write a sentence to explain.

_____

4. It is not good for us to eat nothing but apples. Write a sentence to explain why not. Think of the three important words.

_____

_____

5. Which types of food help us to grow? _____

Which types of food help us to be active? _____

 6. Underline the correct words in the brackets.

My cat eats only fish. Her diet is (**varied / unvaried**). She gets enough

fish every day, so her diet is (**adequate / inadequate**).

7. If a person's daily diet is one slice of bread, one small fish, one spoonful of oil, one bean and one strawberry, then is it:

a varied diet? _____   an adequate diet? _____

 Make a chart to show the five main groups of foods. Write the reason why each food group is good for us (for example, **fats** help us to keep warm). For each group, name two foods that we can buy from shops.

Find the foods

In a supermarket, we can buy many different items.
Tick the items that are foods.

apples ☐	yogurts ☐	potatoes ☐
bananas ☐	tomatoes ☐	rice ☐
brushes ☐	pizzas ☐	spaghetti ☐
cornflakes ☐	coats ☐	sweet potatoes ☐
nails ☐	butter ☐	books ☐
bread ☐	television sets ☐	plantain ☐
fish ☐	pasta ☐	grapefruit ☐
hats ☐	forks ☐	
cakes ☐	newspapers ☐	

Plan your lunch. Choose three or four foods that would make a good lunchtime meal.

_____

SCHOLASTIC DEVELOPING SCIENCE LANGUAGE for Living Things with 8–9 year olds

Hidden words

Find ten words about diets in this wordsearch. Draw a line through each word. Some are horizontal and some are vertical.

p	k	j	v	r	q	w	a
b	a	l	a	n	c	e	d
g	w	x	r	n	u	t	e
r	t	d	i	e	t	x	q
b	o	n	e	t	g	b	u
f	o	o	d	t	u	n	a
h	e	a	l	t	h	y	t
u	y	a	c	t	i	v	e

Write the ten words in alphabetical order.

1. a _ _ _ _ _
2. _ _ _ q _ _ _ _ _
3. _ _ l _ _ _ _ _
4. _ o _ _
5. _ _ _ t
6. _ _ o _
7. _ _ _ _ _ _ y
8. _ u _
9. t _ _ _
10. _ _ r _ _ _

Which word is the name of a part of your body? _____

Which word is an animal? _____

What group of foods does this animal help to provide for us? _____

Why would one of the words be interesting to a squirrel? _____

_____

Types of food

Write each food in the correct column in the table below. Use books, magazines, leaflets, food labels or anything else to help you.

 apple  pasta  cornflakes  pea  sardine

 beef  cabbage  lettuce  banana  carrot

 lamb  potato  trout  vegetable oil  orange

 butter  pear  margarine  salmon 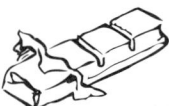 chocolate

 bread  milk  strawberry  rice  cheese

 yam  coconut  kulfi  dumpling  ghee

meat and fish	fats	starches and sugars	vegetables	fruits

SCHOLASTIC DEVELOPING SCIENCE LANGUAGE for Living Things with 8–9 year olds

How many units?

If you look carefully at the labels on tins or packets of food in shops, you will see **nutrition information** like this:

NUTRITION INFORMATION		
Typical Values	Amount per 100g	Amount per serving (200g)
Energy	268kJ/64kcal	536kJ/128kcal
Protein	0.9g	1.7g
Carbohydrate (of which sugars)	7.1g (5.3g)	14.2g (10.7g)
Fat (of which saturates)	3.6g (0.3g)	7.2g (0.5g)
Fibre	0.4g	0.8g
Sodium	0.4g	0.8g
Per Serving (200g) : 128 Calories 7.2g Fat		

To find out more, visit our website at www.heinz.co.uk or write to the address on this can for one of our information leaflets.

Nutritional information from Heinz Cream of Tomato Soup © H J Heinz Company Ltd.

Look at the figure for 'Energy'. This tells us how much **energy** we can use in our bodies, to help us be active, as a result of eating the food. The common **units** for energy are **calories (cal)** and **kilojoules (kJ)**.

This list gives some more energy information:

- One slice of bread provides   470kJ
- One portion of breakfast cereal provides   290kJ
- Seven strands of spaghetti provide   250kJ
- One glass of milk provides   540kJ
- One boiled egg provides   310kJ
- One apple provides   200kJ
- One banana provides   420kJ

Which of these foods provides the most energy? _____

How much energy do two boiled eggs provide? _____

Does a banana provide more energy than an apple? _____

How much more energy can we get from a banana than from an apple? _____

Which food provides the least energy? _____

If you eat one piece of bread, one boiled egg and one apple for lunch, how much energy will you get from your meal?

_____ kJ + _____ kJ + _____ kJ = _____ kJ

What do animals eat?

Fill in the column on the right to show
what each animal usually eats.
Use books or CD-ROMs to help you.

Animal	What does it usually eat?
human	
lion	
dog	
cat	
sparrow	
whale	
rabbit	
goat	
sheep	
cow	
tiger	
elephant	
panda	
monkey	

Ask others in your class or at home what foods they have written for
'human'. List some of them here.

Do you think humans have more foods to choose from than
other animals?

SCHOLASTIC DEVELOPING SCIENCE LANGUAGE for Living Things with 8-9 year olds

Health care

"That is a silly thing to say," said Ruby to Salma. "Your body is not like a bike. How can a body be like a bike?"

"I know they are not exactly the same," said Salma. "A bike doesn't eat or drink like I do."

"So how are they alike?" said Ruby.

"Well, if you don't look after them, they will not keep **healthy**. They might even break down and stop working properly."

"Mmm," said Ruby, "so how do you look after your bike?"

Salma replied: "Well. I wash it to stop dirt getting to the inside. I  dry it so that the rain doesn't rust it. I oil it so that the parts run smoothly – and, most important, I use it so that the different parts keep themselves strong and smooth and… Now it's your turn. How do you look after your body?"

Ruby pondered. "I try to eat the right types of food: not too much, not too little, just enough. I do **exercises** such as walking, running and swimming, and a little special training to **tone up** my muscles. I like to be **active**, but my muscles ache sometimes after too much exercise, and I feel stiff and out of breath for a long time. That's just like your bike, isn't it? Being stiff to move, even though it does not have muscles."

"Yes," said Salma, "so we agree now. To keep our bikes and our bodies healthy, they need to be cared for and used. Our bodies can do many more different kinds of activity than our bikes."

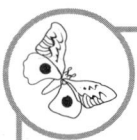

Health care

1. What had Salma just said when this story begins?

_____

2. What does a human do that a bike does not?

_____

3. What must Salma do for her bike and her body?
Circle the correct answer.

　　　　look after them　　　　　　　　leave them alone

4. Name two ways to keep the bike in good condition.

_____　　_____

5. Name two ways to keep the body in good health.

_____　　_____

6. If your muscles are not in good condition, how do you feel after doing exercise?

_____

7. What does 'toning up your muscles' mean?

_____

_____

8. What does Ruby mean when she says that she likes to be **active**?

_____

_____

Name three mechanical things, not including a bike, that do not work well if they are not looked after. Explain what goes wrong and how it can be put right. Draw pictures to illustrate your writing if you wish.

Health care

1. What had Salma just said when the story begins?

_____

2. A human needs to feed. Does a bike need to feed? _____

3. What does Salma have to do for her bike and her body? Draw a circle around one of these statements.

> look after them   leave them alone

4. Name two ways to look after the bike.

_____   _____

5. Name two ways to look after the body.

_____   _____

6. If your muscles are not **toned up** (in good condition) when you have done exercise, how do they feel?

Do they ache? _____ Do they squeak? _____ Do they rust? _____

7. When your muscles have been **toned up**, which of these words describes them? Tick the best words.

stiff ☐   supple ☐   strong ☐   weak ☐

8. Ruby says that she likes to be **active**. What does she mean? Tick the best sentences.

She likes to play running games. ☐   She likes to dance. ☐

She likes to swim. ☐   She likes to sleep. ☐

Name three mechanical things, not including a bike, that do not work well if they are not looked after. Explain what goes wrong and how it can be put right. Draw pictures to illustrate your writing if you wish.

Feel the pulse

Zora and her brother Mark were walking home from school and talking about what they had been doing.

Mark: We drew charts in maths.
Zora: We looked at the **pulse** in science.
Mark: What's the pulse?
Zora: It's the beat that you feel when you put your fingers on your wrist, like this. It tells you what your heart is doing.
Mark: That's silly. How can your wrist know what your heart is doing? They're too far apart.
Zora: Your heart pumps blood all around your body. The beats in your wrist show your heart pumping the blood into your arm. If you count them, you know how fast your heart is beating.
Mark: Let's try it. You count my pulse and I'll count yours. We'll write them down on a piece of paper, then you can show them to your teacher.

This is the paper that Zora showed to her teacher.

o Mark 14
o Zora 75

What do you think the teacher said when she saw the paper? Write three things that she might have said. Think about fair testing and measuring time.

1. _____

_____

2. _____

_____

3. _____

_____

SCHOLASTIC DEVELOPING SCIENCE LANGUAGE for *Living Things* with 8–9 year olds

Feel the pulse

The next day, Mark and Zora walked home from school again. They decided to have a race for the last part. When they reached home, they stopped at the gate to rest. They were breathing heavily.

Mark: I can feel a thumping in my chest. It's my pulse going fast.
Zora: No, it's not. It's your heart beating faster because you've been running. But your pulse will be faster too.
Mark: OK. Let's measure our pulses.

This is what they found.

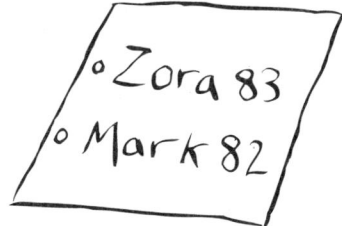

Zora 83
Mark 82

Zora: They're higher than yesterday, when we were only walking. Will they come down again?
Mark: I hope so. I don't like this fast thumping.
Zora: Maybe we can try an experiment with our friends in school tomorrow.
Mark: That's a great idea. We'll write a plan and ask if we can try it out. Can I use my maths charts?
Zora: Yes, that'll be good. What should we put in our plan?

What sort of plan should Zora and Mark write? Can you help them?

On another sheet of paper, write a set of about five instructions to help Mark and Zora investigate how doing an exercise such as running or jumping changes the pulse rate. Remember to say how they should use a maths chart.

Body and brain exercises

Do the following activities help to **tone up** body muscles?
Write 'Yes' or 'No' under each activity.

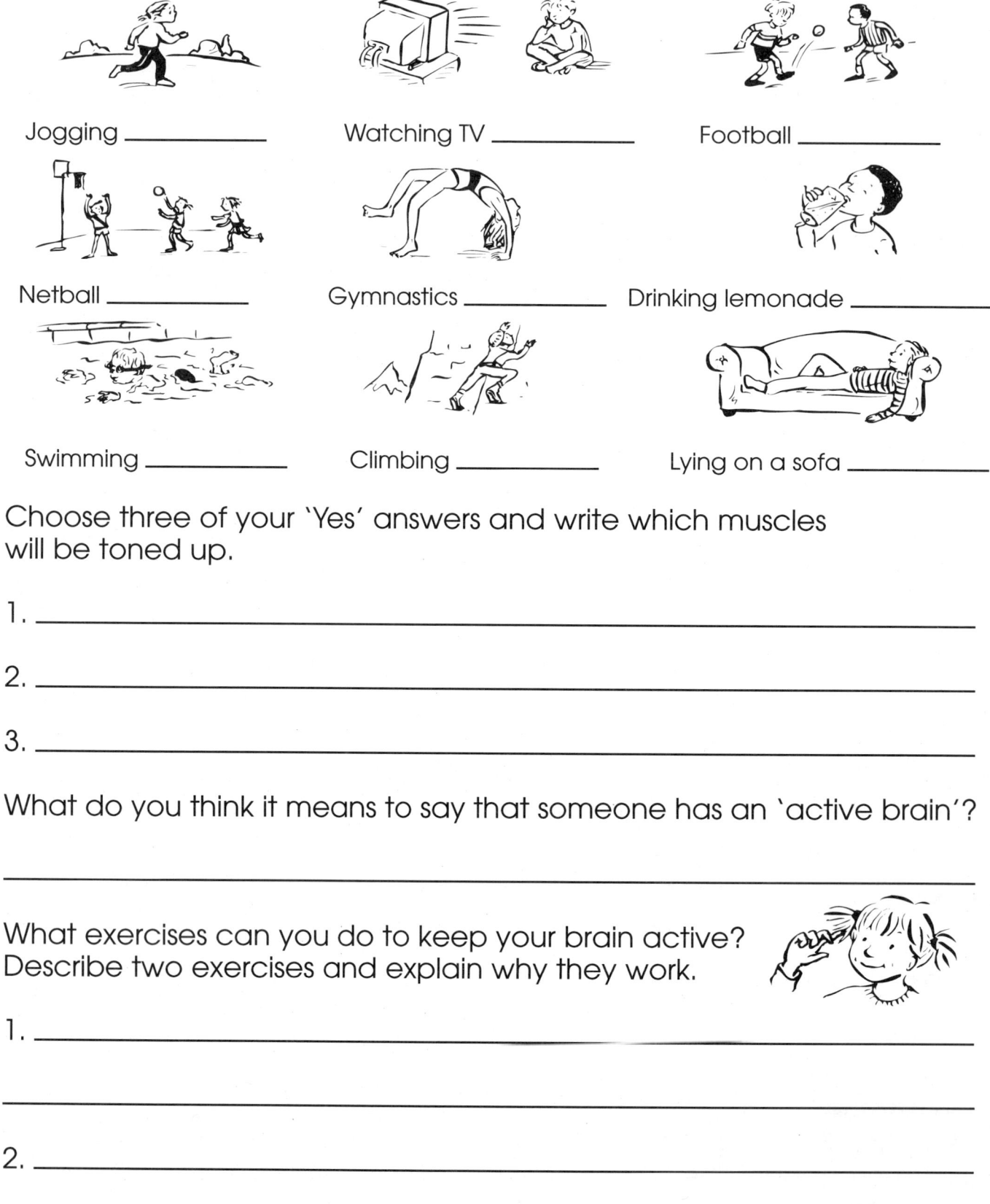

Jogging _____   Watching TV _____   Football _____

Netball _____   Gymnastics _____   Drinking lemonade _____

Swimming _____   Climbing _____   Lying on a sofa _____

Choose three of your 'Yes' answers and write which muscles
will be toned up.

1. _____

2. _____

3. _____

What do you think it means to say that someone has an 'active brain'?

_____

What exercises can you do to keep your brain active?
Describe two exercises and explain why they work.

1. _____

_____

2. _____

_____

▲ SCHOLASTIC   DEVELOPING SCIENCE LANGUAGE for *Living Things* with 8–9 year olds

Plant systems

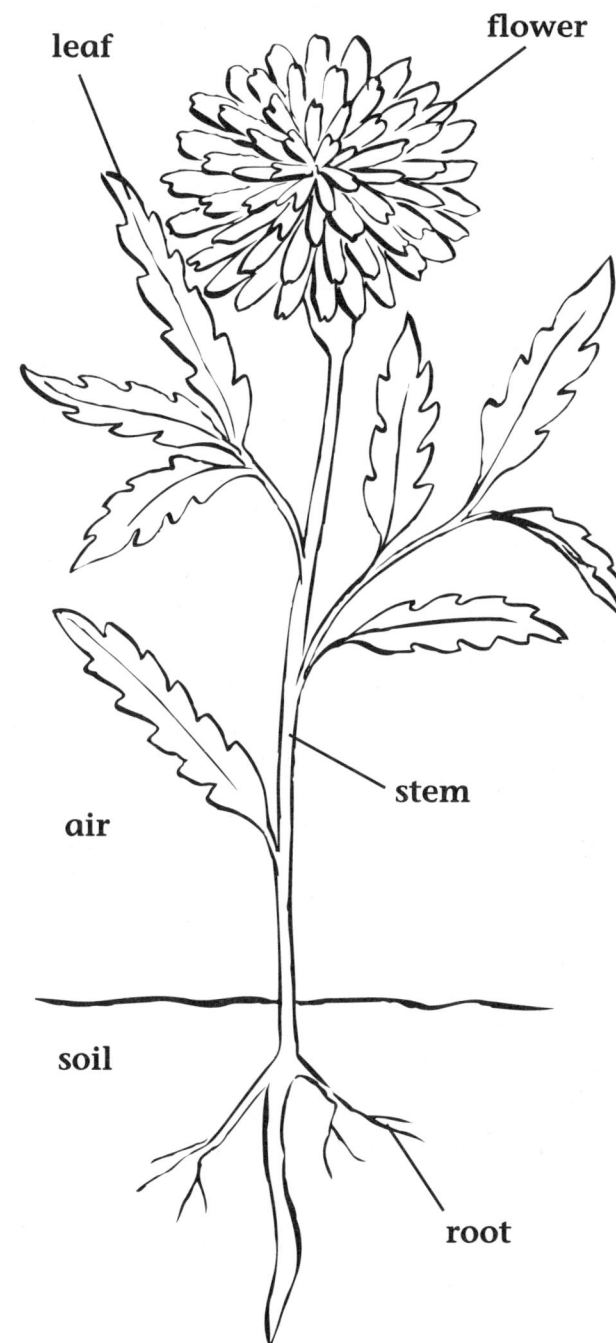

leaf
flower
air
stem
soil
root

Most green plants have **roots, stems, leaves** and **flowers**.

Roots are usually white and grow in the soil. They anchor the plant and take in **water** from the soil.

The **stem** is the main **stalk** above the ground. It holds up the leaves and flowers, and carries water from the roots.

Leaves are green, flat and thin. Leaves capture **sunlight** to make the plant's food. They also **exchange gases** with the air outside the plant.

Flowers are often brightly **coloured** and **scented** to help them attract insects, which help the plants to **reproduce**. This means to make new plants.

Humans grow many plants for food. We eat root vegetables (such as carrots, potatoes and turnips). We eat leaves (such as lettuce and cabbage). Other leaves, such as mint and parsley, are used to give our food **flavour**. We also eat stems (such as celery and rhubarb) and flowers (such as broccoli and nasturtiums).

Plant systems

1. Write the name of the part of a plant that:

holds up flowers _____ anchors the plant _____

takes in water _____.

2. Write the name of: a root vegetable _____

a leaf to flavour food _____ a stem that we eat_____.

3. From where do plants get:

(a) light? _____ **(b)** gases? _____ **(c)** water? _____

4. Which part of a plant reproduces? _____

What does a plant do when it reproduces? _____

_____

5. Flowers from a shop can be put in a vase with some water.
Even so, they soon die. Explain why this is.

_____

6. Underline the correct word or words to complete each sentence.
Sunlight is used in **roots / leaves**. Roots are usually **white / green**.
Plants need sunlight to make **food / water**. Stems usually grow **above the
soil / in the soil**.

7. In which part of the day are plants not able to make food? Why?

_____

_____

Find some old magazines that have pictures of plants. Cut out ten
different pictures. On each picture, label the root, stem, leaf and flower
if you can see them. Write the name of the plant.
Now choose one of the plants and write a factfile about it.

Plant systems

1. Draw a line under the **correct** word at the end of each sentence. The part of a plant that grows above ground is the **root / stem**. Flowers are held up by the **stem / root**. Plants are anchored by **stems / roots**. Water is taken from the soil by the **stem / leaf / root**.

2. Write the names of: **(a)** a root vegetable _____

(b) a leaf to flavour food _____

(c) a stem that we eat _____

3. From where do plants get:

(a) light? _____ **(b)** gases? _____ **(c)** water? _____

4. Cut flowers kept in water in a jar soon die. Why?

_____

5. What helps a flower to make new plants? Tick the **correct** boxes.

They are brightly coloured. ☐   They grow on stems. ☐

They capture sunlight. ☐   They have scent. ☐

6. Cross out the **incorrect** word at the end of each sentence. The green parts of plants are **roots / leaves**. To make food, plants need **sunlight / snow**. Sunlight is used in the **roots / leaves**.

7. Why are plants not able to make food at night? _____

_____

Find some old magazines that have pictures of plants. Cut out ten different pictures. On each picture, label the root, stem, leaf and flower if you can see them. Write the name of the plant.
Now choose one of the plants and write a factfile about it.

Which part does what job?

Draw a line to match each part of a plant to the job that it does.

leaf	anchors the plant in the ground
flower	holds the plant upright above ground
root	makes new plants
stem	makes food

Do you think the word 'anchor' is a good word to describe one of the jobs of the root? Explain why you think that. You may need to look in a dictionary for the meaning of the word 'anchor'.

_____

_____

Look in a dictionary to find other meanings of the word 'root'. Write one of these meanings here, and explain why 'root' is a good word to use for it.

_____

_____

This picture shows what some vandals have done to plants in a park.

Explain to your friend or teacher why these plants will probably die. What could the vandals do to make the park look how it did before?

The hidden plant

Use the clues to find the five missing words. The name of a plant is hidden in the grey boxes.

1.
2.
3.
4.
5.

Clues

1. A living thing is either an animal or a _ _ _ _ _.
2. Roots _ _ _ _ _ _ a plant in the ground.
3. This plant will sting you, but you can make a drink from it.
4. Leaves use this to make food.
5. Do humans eat plants?

What is the name of the hidden plant? _____

Look in books or magazines, or on a CD-ROM, for pictures of this plant. Look carefully at a picture, then make your own drawing of the plant in the box on the left.

On another sheet, write a letter to a friend about this plant. Describe the shape of the petals and the leaves. Say how many petals there are. Describe anything else you notice.

COVER THESE INSTRUCTIONS WHEN PHOTOCOPYING.

Notes for teachers
The answers are as follows: 1. plant 2. anchor 3. nettle 4. sunlight 5. yes.
The hidden plant name is 'pansy'.

Plant parts that we eat

Choose the correct words from the box to fill the spaces in the passage below.

grass	roots	white	sunlight	air	carrots
lettuce	green	flowers	stems	food	rhubarb

Plants are made up of stems, _____, leaves and _____. Leaves are usually _____ in colour and grow on _____ above ground. They capture _____ to make _____. We eat parts of plants. We eat leaves such as _____ and roots such as _____.

Complete this table. Don't use any plant names from the box above.

Roots that humans eat	Stems that humans eat	Leaves that humans eat

Write a sentence about one food from each column: one root, one stem and one leaf. You might like to say where the foods grow, in which time of year they grow at, or how they should be cooked.

1. _____

_____

2. _____

_____

3. _____

_____

Look very closely

This activity helps you to learn a skill that is very important in science: looking closely at something and making an accurate drawing of what you see.

Look very carefully at all the parts of the plant. Make a drawing of the plant on this page. Make sure that the parts are in the correct **proportion** and in the correct places. Colour your drawing like the plant. Write the name of the plant on the label.

Write three questions to ask your friends about your drawing. For example, you might want to ask them where the **leaves** are, or how many **petals** the flower has. Remember to ask hard questions as well as easy ones. Make sure you know the answers!

1. _____

_____

2. _____

_____

3. _____

_____

COVER THESE INSTRUCTIONS WHEN PHOTOCOPYING.

Note for teachers
You need to provide one plant for every four children. Any plant will do, but it is best to choose one with a relatively simple structure. The plant should have leaves and at least one flower. It could be presented in a pot.

A plant dictionary

The letters of these six important words about plants have been mixed up.

relfwo   nshilgut   elgevateb   veslae   ostro   emst

Rearrange the letters to make words. Write the words here in alphabetical order.

_ _ _ _ _ _   _ _ _ _ _ _   _ _ _ _ _

_ _ _ _   _ _ _ _ _ _ _ _   _ _ _ _ _ _ _ _

Use these words to make a plant dictionary. Write the six words in alphabetical order in circles 1 to 6 on this flower picture. Under each word, write a definition of it.

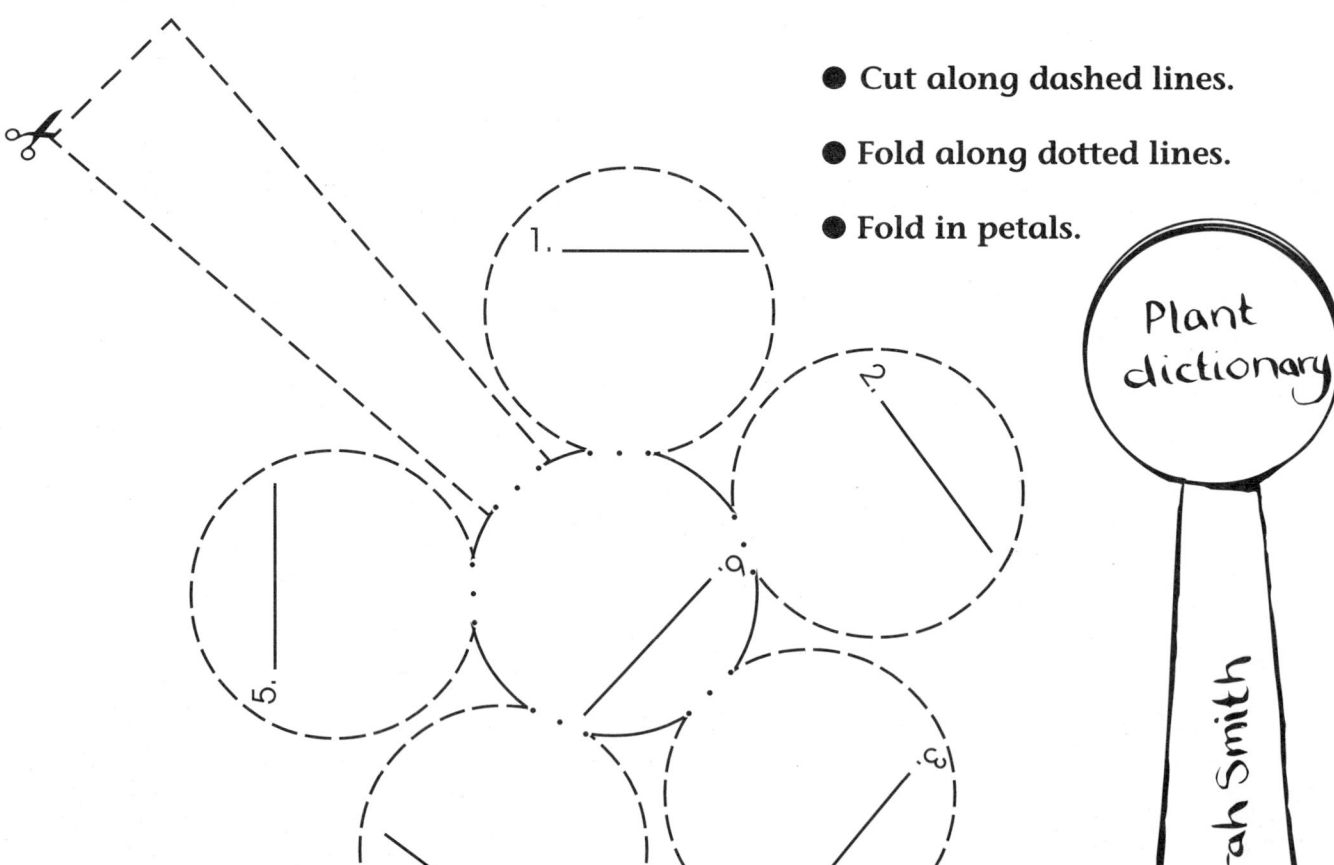

● **Cut along dashed lines.**

● **Fold along dotted lines.**

● **Fold in petals.**

Write your name on the stem. Fold up the petals so your dictionary can be read in alphabetical order. Write 'Plant dictionary' on the back of petal 1.

■SCHOLASTIC   DEVELOPING SCIENCE LANGUAGE for *Living Things* with 8–9 year olds

Plants and their needs

For plants to grow well and be healthy, they need **water** and **minerals** from the soil, **sunlight**, **air**, **warmth** and **space** to grow.

If a plant does not have enough water, it soon becomes limp and starts to droop. If it is left too long without water, a plant may lose its leaves and eventually die. The roots of a plant take in water and dissolved **minerals** from the soil. The plant grows best when there is plenty of space for its roots to spread out.

Plants grow well when they are able to use sunlight to make **food** in their leaves. A special substance that gives the leaves their green colour is necessary for this. Plants that are left in the dark have pale or yellow leaves, because without light the leaves do not make the special substance.

Plants do not usually grow well in winter, because it is too cold. Sometimes gardeners grow plants in **greenhouses**, which have lots of windows or panels made of glass or clear plastic. Sunlight passes through the panels and warms the soil and air around the plants. The

greenhouses also protect the plants from the strong winds, heavy rain, frost and snow that often come in winter. With this protection, the plants can stay alive and grow.

Plants and their needs

1. Name three things that plants need to grow well.

1. _____   2. _____   3. _____

2. Which part of a plant:

a) takes in minerals from the soil? _____

b) makes food? _____

3. Would a plant grow well in a small bucket filled with sand? Explain your answer by writing a sentence.

_____

4. What happens to a plant that does not get enough light?

_____

5. Rachel put two plastic sheets near to each other on a patch of grass. One sheet was black and the other was clear.

Two weeks later the grass under the black sheet was yellow. Why was this?

_____

6. Why do you think greenhouses are called greenhouses, when usually the glass has no colour? You might need to look in a book to find the answer.

_____

Ranjit noticed that more grass and small plants grew in the space between two trees than under the trees.
Explain why this has happened. Make a labelled diagram of the trees and the other plants to help with your explanation.

Plants and their needs

1. Name three things that plants need to grow well.

1. _____   2. _____   3. _____

2. Which part of a plant:

a) takes in minerals from the soil? _____

b) makes food? _____

c) needs space to spread out? _____

3. Jake put a plant to grow in soil in a bucket that has very small holes in its side. Draw what you think will happen. Explain why you think that will happen.

_____

4. Does a plant get its water from the soil? _____

Does a plant need light to make its food? _____

5. Rachel put two plastic sheets near to each other on a patch of grass. One sheet was black and the other was clear.

Two weeks later the grass under the black sheet was yellow. Why was this?

_____

6. Greenhouses are not green, so why do they have this name? You might need to look in a book to find the answer.

_____

 Ranjit noticed that more grass and small plants were growing in the space between two trees than under the trees.
Explain why this has happened. Make a labelled diagram of the trees and the other plants to help you explain.

True or false?

Tick the correct box to show whether each sentence is true or false.

	True	False
1. Roots are green.		
2. Leaves need light to make food.		
3. Plants without water may lose their leaves.		
4. Plants take water from the air.		
5. Plants do not need minerals to grow well.		

Write five more sentences to try on your friends. Make sure that you know whether each sentence is true or false!

	True	False
1. _____ _____		
2. _____ _____		
3. _____ _____		
4. _____ _____		
5. _____ _____		

SCHOLASTIC DEVELOPING SCIENCE LANGUAGE for *Living Things* with 8–9 year olds

Healthy and unhealthy

A   B

Which of these plants is healthy?_____

Write down four features that show it is a healthy plant.

1. _____

2. _____

3. _____

4. _____

What do you think might have caused the other plant to be unhealthy? Describe two possible causes. You can use words from the box to help you.

plants	water	enough	soil	sunlight	dry	wet

1. _____

2. _____

If someone asked you to make the unhealthy plant healthy again, what is the first thing that you would try? Explain why.

I would try _____

because I noticed that _____

Gardener's world

Choose a word from the box to complete each sentence. You can use each word one, more than one or not at all.

| plants | air | minerals | water | light | animals | soil |

1. Plants provide food for some _____.

2. A drooping plant is not getting enough _____.

3. Roots take in water and _____.

4. Yellow or pale leaves mean
that a plant is not getting enough _____.

5. Green leaves take in _____ and _____ to make food
for the plant.

6. Plants usually grow well when they are in rich _____.

Write three instructions for gardeners to help them grow healthy plants. Think about what plants need to grow well.

1. _____

2. _____

3. _____

What living things might harm plants in a garden? Write two more instructions for gardeners to help them protect their plants from harm.

1. _____

_____

2. _____

_____

■■ SCHOLASTIC DEVELOPING SCIENCE LANGUAGE for Living Things with 8–9 year olds

Dr Gardener

Look at these pictures of unhealthy plants. If you were a gardener, how would you treat the plants to make them healthy again? Give a reason for your answer each time.

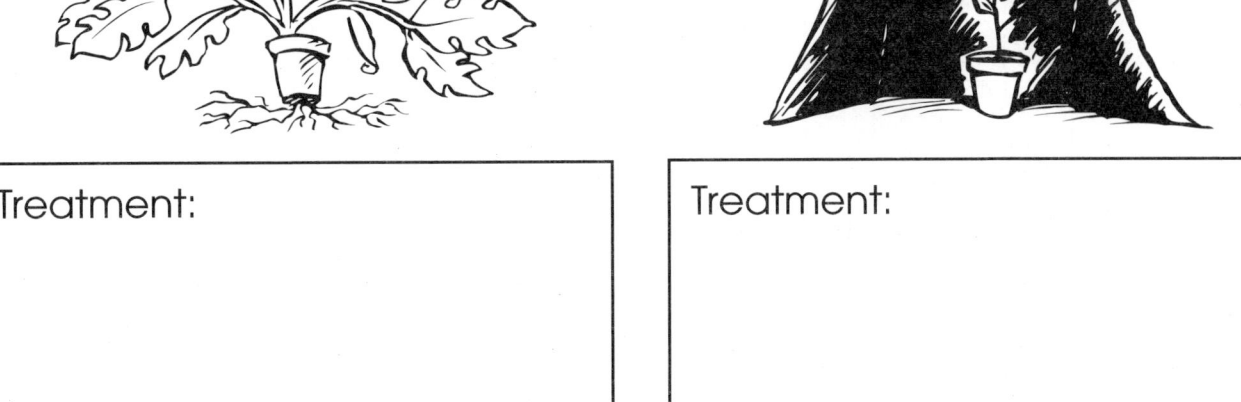

| Treatment: |
| Reason: |

| Treatment: |
| Reason: |

| Treatment: |
| Reason: |

| Treatment: |
| Reason: |

Good for plants and good for humans?

Can you find six words that are the names of things that plants need?
Draw a line through each one that you find. The words may be
horizontal, vertical or even diagonal.

M	A	B	Y	O	F	T
I	E	B	C	D	H	X
N	C	E	R	G	L	Z
E	A	M	I	N	Y	U
R	P	L	O	R	I	A
A	S	W	A	T	E	R
L	S	G	K	R	I	V
S	W	A	R	M	T	H

Write the six words here in alphabetical order.

1. _____   4. _____

2. _____   5. _____

3. _____   6. _____

These words say what plants need to be healthy. Do they also say what
humans need to be healthy? For each word, write a sentence to say
whether humans need the same thing.

1. _____

2. _____

3. _____

4. _____

5. _____

6. _____

   SCHOLASTIC DEVELOPING SCIENCE LANGUAGE for Living Things with 8-9 year olds

The human skeleton

When we press our **fingers** over our **skin**, we can feel that some parts under the skin are soft and some are hard. The hard parts are our **bones**. We have many bones of different shapes and sizes. They are all connected together by **joints** to make one structure, called the **skeleton**.

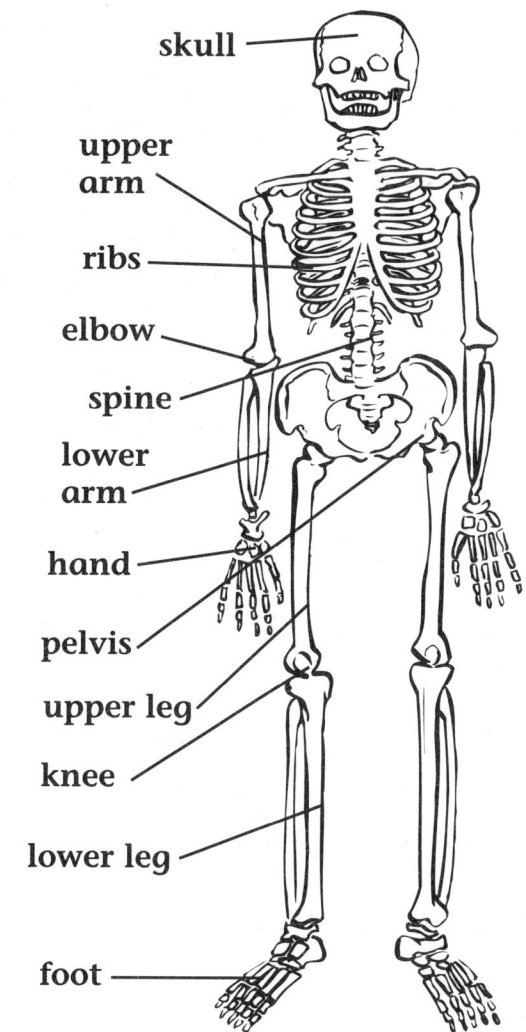

skull
upper arm
ribs
elbow
spine
lower arm
hand
pelvis
upper leg
knee
lower leg
foot

Our skeleton helps us to stand up, and helps us to move in different ways.

● The **spine**, or **backbone**, makes the human body able to stand upright. Without it, the top part of the body would flop down.

● The **skull** is important because it protects the **brain**.

● The **ribs** are important because they protect the **lungs** and **heart**.

Although our bones are hard, they can break easily if we fall or something hits us. With care, most broken bones mend and become strong again.

As we develop from babies to adults, we grow bigger because our bones are growing all the time. A teenager is taller than a baby, because the teenager's backbone and legs have grown longer.

The human skeleton

1. Which parts of our body feel hard under our **skin**? _____

2. We have many **bones**. Name three of them.

_____   _____   _____

3. Complete this sentence: Our bones are different from each other

because _____

4. The human skeleton is **symmetrical**. Name two bones that are very

similar to each other. _____

Explain why it is important that they are similar. _____

_____

5. What two main jobs do our **skeletons** have? _____

_____

6. Different bones have different jobs to do. Write about what
two of them do.

a) _____

b) _____

7. What are animals that have **backbones** called?

8. Name one thing that would be difficult to do if your arms could not

bend at the elbow. _____

9. Name two **invertebrates**. _____   _____

How many bones do you think there are in your body? Make a list of
names of bones, then find out how many there are of each type. For
example: skull (1), upper arms (2). There are many little bones in our
hands and feet, but you do not need to count them all.

          SCHOLASTIC   DEVELOPING SCIENCE LANGUAGE for Living Things with 8–9 year olds

The human skeleton

1. What are the hard parts under our **skin** called? _____

2. Do we have many **bones** or only a few? _____

Name three bones.

_____   _____   _____

3. Our bones are not all the same size. In what other way are they

different? _____

4. Some bones are **similar** to each other. In which parts of the human body are they found? Remember that the human body is **symmetrical**.

_____ and _____

5. What two main jobs do our **skeletons** have?

_____

_____

6. What job does the **skull** do? Write a sentence.

_____

7. Can you think of one action that would be difficult to do if your arms were fixed straight? (Think of mealtimes.)

_____

8. Complete this sentence.
Animals with backbones are called v_____

9. Name two **invertebrates**. _____ and _____

How many bones do you think there are in your body? Make a list of names of bones, then find out how many there are of each type. For example: skull (1), upper arms (2). There are many little bones in the hands and feet, but you do not need to count them all.

Buried bones

Can you find eight words about the human skeleton in this wordsearch? Draw a line through each word you find.

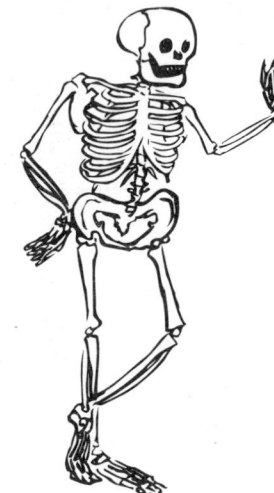

A	F	Z	E	S	K	U	L	L	Y
B	P	U	I	K	H	V	G	X	G
V	E	R	T	E	B	R	A	T	E
J	L	C	T	L	D	I	W	A	L
P	V	K	I	E	D	B	H	F	B
R	I	Q	L	T	B	S	E	C	O
O	S	J	B	O	N	E	S	M	W
S	K	N	K	N	E	E	S	L	S

Write the eight words here in alphabetical order.

1. _____   5. _____

2. _____   6. _____

3. _____   7. _____

4. _____   8. _____

If your words are in the right order, the words at numbers 2, 3, 5 and 7 are bones with particular **functions** (this means work or jobs) to do. Fill in this table to name each of these types of bone and say what it does.

Bone number	Name	Main function
2		
3		
5		
7		

▬SCHOLASTIC DEVELOPING SCIENCE LANGUAGE for *Living Things* with 8–9 year olds

Skeleton key

Label this diagram of the human skeleton. The words you need are in the box below. You may need to look in books or ask an adult to help with some of the words.

skull	spine	elbow	knee	jaw	collarbone	femur
fibula	tibia	wrist	rib	ankle	knuckle	pelvis

Which of the four words are **joints**? Think about how your body moves.

_____   _____

_____   _____

Word tower

Write the solutions to the eight clues in the boxes by the numbers.
You may need to look in books for help.

Clues

1. This protects your brain.
2. This joins your leg to your foot.
3. This bone is the upper part of your leg.
4. This is where your arm bends.
5. This is where your legs join the rest of your body.
6. This bone is in the lower part of your leg.
7. These are the hard things that you feel through your skin.
8. This is where your leg bends.

What is the long word down the middle of your word tower?

On another sheet of paper, write an acrostic poem with the letters of
this word. Write lines in your poem to help your friends understand how
the mystery word works.

SCHOLASTIC DEVELOPING SCIENCE LANGUAGE for Living Things with 8–9 year olds

Skeleton card game

Sentence starters	Sentence endings
* My ribs help	they can breck if I fall.
My knee joins	to protect my lungs.
I am a vertebrate	the upper and lower parts of my leg.
My skeleton is made	because I have a backbone.
My wrist is where	of very many bones.
My skull helps	my lower arm joins my hand.
I have two knuckles	to protect my brain.
My elbow is where	in each of my fingers.
My chin is the front	the upper and lower parts of my arm join.
My ankle is where	of my lower jawbone.
There are many bones	my lower leg joins my foot.
Worms are invertebrates	in my feet.
Another name for skull	because they do not have backbones.
Another name for backbone	is cranium.
Although my bones are hard	is spine.

COVER THESE INSTRUCTIONS WHEN PHOTOCOPYING.

Teacher instructions

Photocopy onto card. Cut along the dotted lines. Fold each card in half along the solid line, with the text on the outside, and fasten with adhesive tape.

If you are working with a small group, give each child a card. If you are working with the whole class, share the cards out one between two or three. All the cards must be given out.

The child (or group) with the card marked * reads the sentence starter aloud. The child (or group) with the correct ending to that sentence reads it out, then reads out the starter on the back. This goes on until the first child (or group) has read out the ending on the first card.

1

Make a skeleton puppet

You can use the parts drawn below to make a skeleton puppet. The picture opposite shows the whole puppet when it is finished.
But I've lost the instructions!

Please help by writing some new instructions. Cut out the puppet parts and stick them on to pieces of cardboard.
Decide how you can join the different parts together, so that the puppet will move almost in the way that a skeleton moves. Use sticky tape or paper fasteners.

Now write some instructions on another piece of paper for me to follow to make a puppet. Remember to list all the things I'll need.

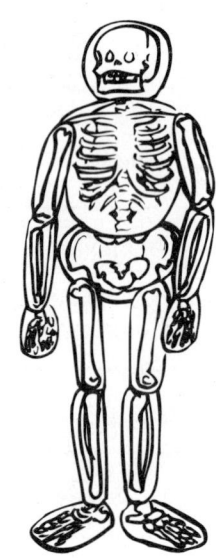

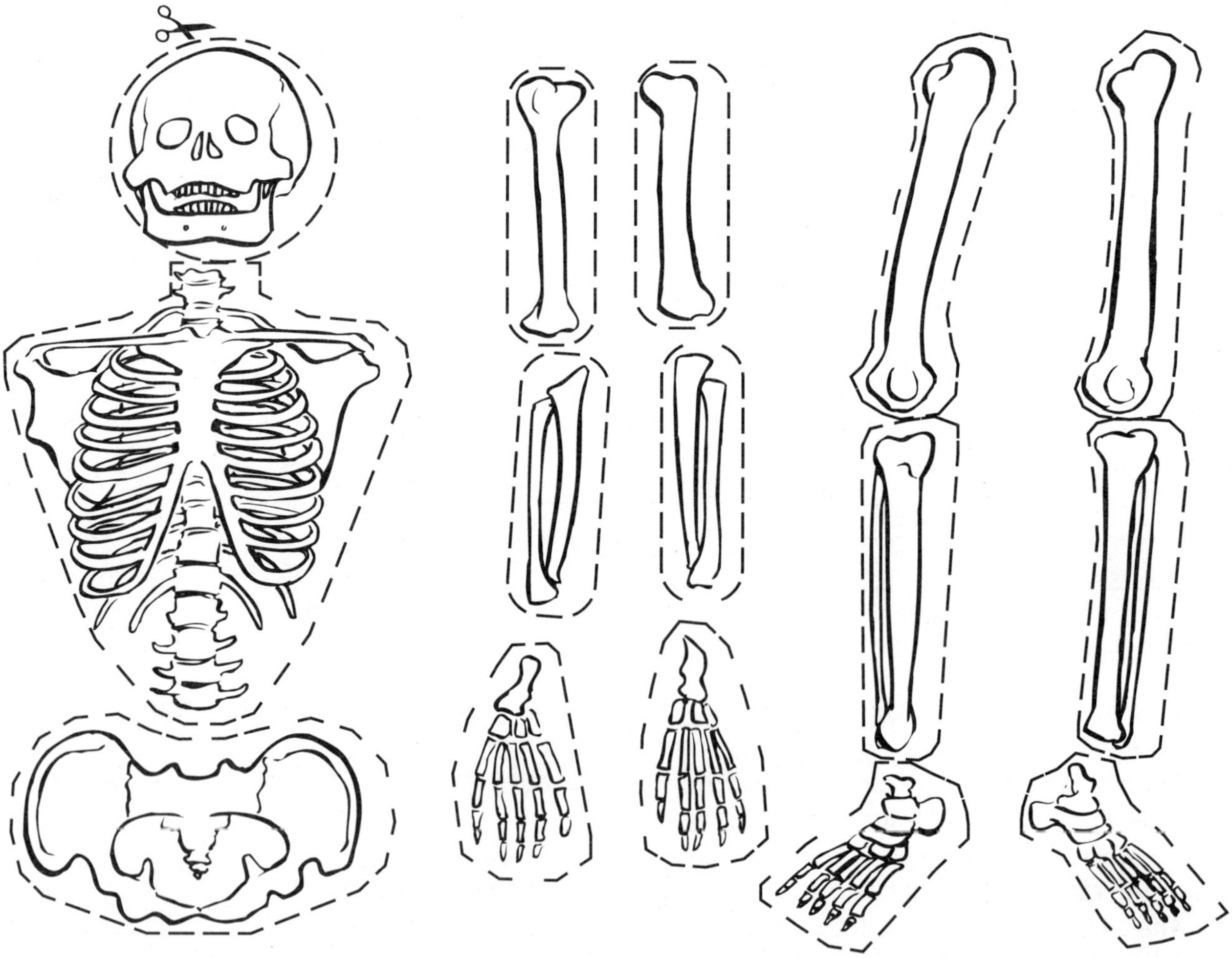

Make a skeleton puppet

When you have made the puppet, look carefully at how it moves.

How are the ways that it moves different from the ways a real human skeleton moves?

Write five differences. Start each sentence with: 'A real skeleton can…'.

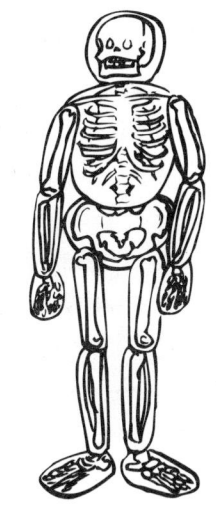

1. A real skeleton can _____

_____

_____

2. _____

_____

_____

3. _____

_____

_____

4. _____

_____

_____

5. _____

_____

_____

How our bones protect us

The human skeleton protects some of our most important **organs**. Organs are parts of the human body that are necessary for us to stay alive and healthy. They are usually soft and can easily be damaged.

The organs hidden inside our bodies are known as **internal** organs. 'Internal' means inside. Here are the names of some of our internal organs:

| brain | heart | lungs | liver | spinal cord |

1. Extend the line from each label on the left of this picture to show where that organ is in the body. You could draw in the organ too.
2. Write the name of the part of the skeleton that protects each of these organs.

Which organ?

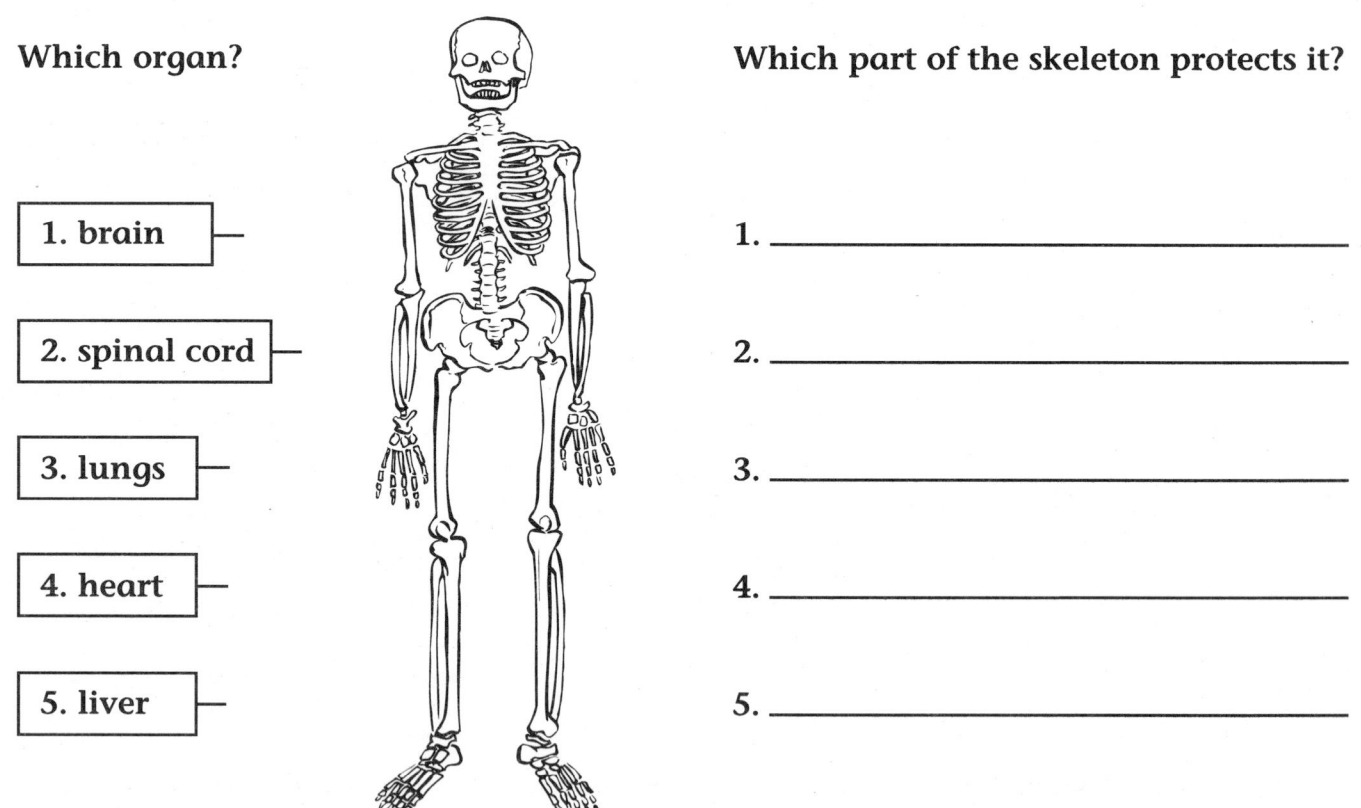

1. brain

2. spinal cord

3. lungs

4. heart

5. liver

Which part of the skeleton protects it?

1. _____

2. _____

3. _____

4. _____

5. _____

3. Draw an arrow from each skeleton part label to show where it is on the skeleton.

4. Find out about the spine. Write a report on how the spine helps us.

SCHOLASTIC DEVELOPING SCIENCE LANGUAGE for *Living Things* with 8-9 year olds

Other animal skeletons

Every living thing is either a **plant** or an **animal**. Humans are animals. The animal group also includes whales, dolphins, gerbils, owls, lizards, frogs, sharks, trout, centipedes, ants, scorpions, jellyfish and many more.

All animals are **similar** to each other in important ways. They all **feed**, **move** from place to place, **reproduce**, get rid of **waste** material from their bodies, **grow** and can **sense** the world around them.

Animals may also be **different** from each other. One important difference is that some animals have a **backbone** and others do not. Animals that have a backbone are called **vertebrates**. Animals that do not have a backbone are called **invertebrates**.

Humans are vertebrates. So are other animals such as dogs, cats, sparrows, lizards, snakes, newts, pikes and eels. The invertebrate group includes wasps, beetles, spiders, snails, worms and starfish.

There is another important difference between the **skeletons** of vertebrates and those of invertebrates. Vertebrates have skeletons inside their bodies, usually made of **bone**. Many invertebrates, such as beetles, crabs, lobsters and snails, have skeletons that are outside their bodies (like a suit of armour). If this **external** skeleton is hard and brittle, it is called a **shell**. Snails have shells.

Some invertebrates, such as worms, have no skeleton at all. What keeps the worm's body firm is an **internal** tube filled with water.

Other animal skeletons

1. Name four members of the **animal** group of living things.

_____

2. Name three **similarities** between animals.

_____

3. What is a **vertebrate**? Write a sentence.

_____

4. Is a jellyfish an **invertebrate**?
Write a sentence to explain your answer.

_____

5. Draw a line through the odd one out.

cow   gerbil   worm   mackerel   tortoise

Why is it the odd one out? _____

6. Explain what the main difference is between the skeleton of a horse and the skeleton of a crab.

_____

_____

Sort these animals into 'vertebrates' and 'invertebrates' and list them separately. You can decide how to show the lists.

human, horse, cat, eagle, shark, worm, cod, crocodile, starfish, elephant, rabbit, sparrow, fly, snail, insect, pterodactyl, jellyfish, centipede, spider, dolphin, penguin, snake, crab, lobster

Now sort the whole set of animals into two different groups. Show the two groups separately, and write captions or labels to explain what is special about each group.

SCHOLASTIC DEVELOPING SCIENCE LANGUAGE for Living Things with 8–9 year olds

Other animal skeletons

 1. Name four members of the **animal** group of living things.

_____

2. Name three **similarities** between animals.

_____

3. Cross out the incorrect words in this sentence.

A vertebrate **has / does not have** a backbone.

 4. Draw a line through the odd one out.

cow	gerbil	goldfish	tortoise	worm

Why is it the odd one out? _____

5. A jellyfish is an **invertebrate**. What does this mean?

It means that _____

6. The skeleton of a horse is inside its body. It is made of bones. Explain how the skeleton of a crab is different.

_____

Sort these animals into **vertebrates** and **invertebrates**. Show them separately in any way you wish.

human	spider	insect	shark	pterodactyl
jellyfish	crab	horse	elephant	

Now sort them all into two different groups. Write words to tell your teacher or a friend what is special about each group.

Like and unlike

These pictures show the skeleton of a horse and the skeleton of a human.

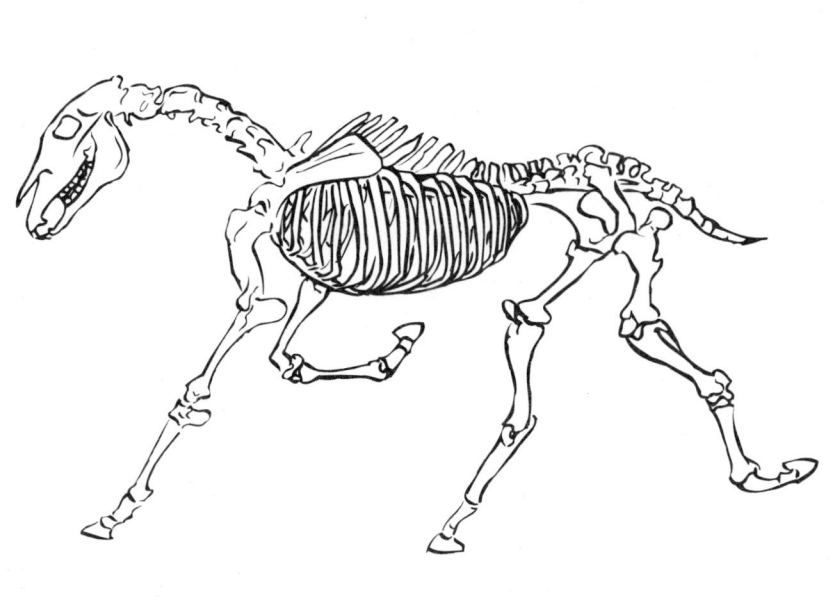
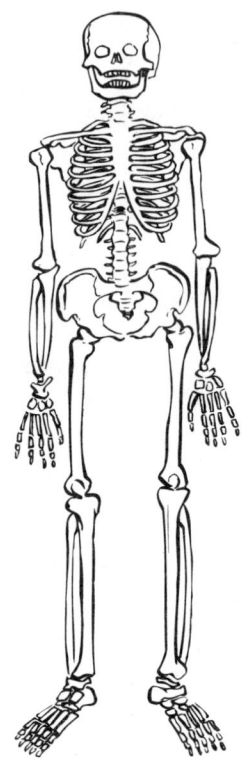

Use the table to make two lists. First, list the ways in which the horse and human skeletons are **similar**. Second, list the ways in which the skeleton of the horse is **different** from the skeleton of the human.

Similarities	Differences

Whose skeleton is it?

Under each picture of a skeleton, write the name of the animal it belongs to.

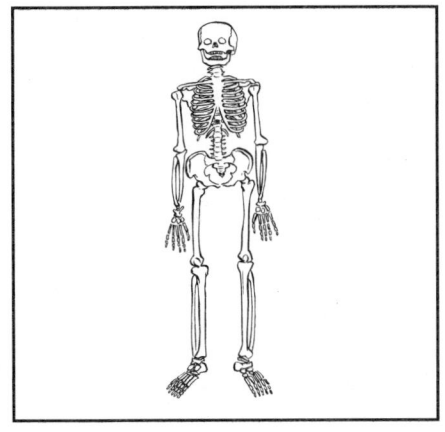

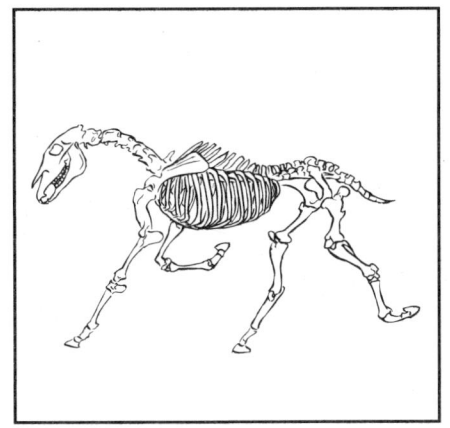

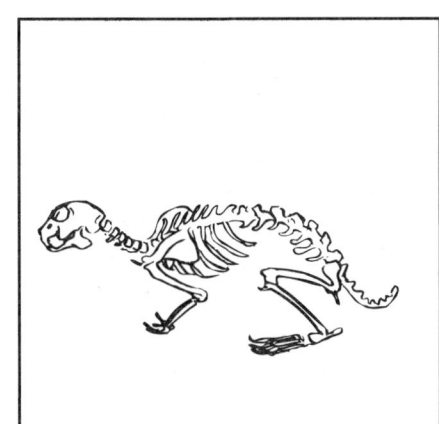

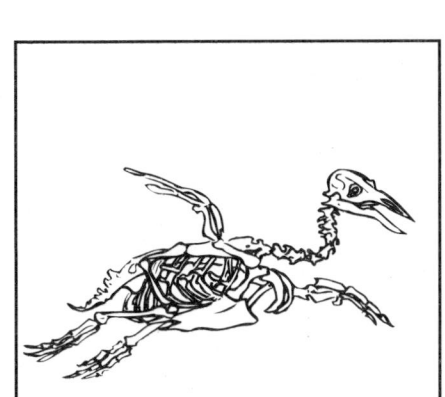

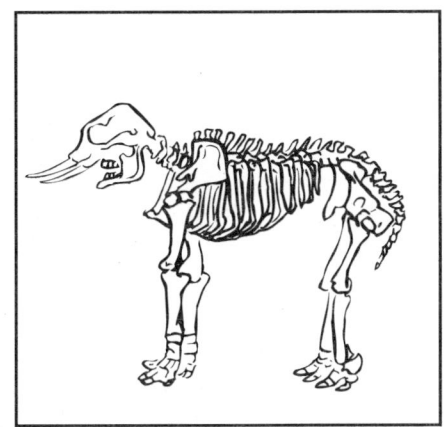

Write the names of the animals in order of their size, starting with the smallest.

_____

| penguin | elephant | human | dinosaur | horse | rabbit |

┌─ **COVER THESE INSTRUCTIONS WHEN PHOTOCOPYING.** ─┐

Note for teacher
The answers in the box can be given to the children to make the task easier, or covered up before the page is photocopied.

Odd one out

Look at these sets of pictures. Which is the odd one out in each set?
Think about skeletons. Write a sentence to explain each answer.

| cat | snail | worm | bird |

Odd one out: _____

Reason: _____

| goldfish | crocodile | crab | cod |

Odd one out: _____

Reason: _____

| lobster | crab | bee | snail |

Odd one out: _____

Reason: _____

| jellyfish | starfish | sea anemone | frog |

Odd one out: _____

Reason: _____

SCHOLASTIC DEVELOPING SCIENCE LANGUAGE for Living Things with 8–9 year olds

Human muscles

Our **skeletons** enable us to stand up straight. The **joints** in our body, such as the **shoulder**, **elbow**, **knee** and **wrist**, enable different parts of our bodies to move. Without the joints, the skeleton would be rigid like a post.

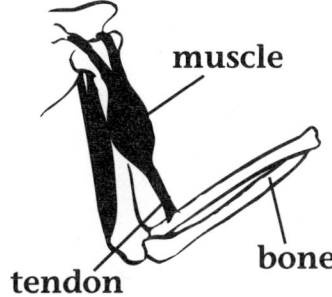

But the bones cannot move by themselves. What makes them move? The answer is **muscles**. These are made of soft, but strong, body **tissue**. Muscles are attached to bones by a very strong body tissue called **tendons**.

There are different kinds of muscles in our bodies. The kind of muscles that make our bodies move work by pulling on bones. They cannot push. Muscles pull when they **contract**, which means that they become shorter. When they do not need to pull, they **relax** and return to their normal length. Our **brains** send messages to our muscles to tell them what we want them to do.

The muscles in our arms and legs work in pairs:

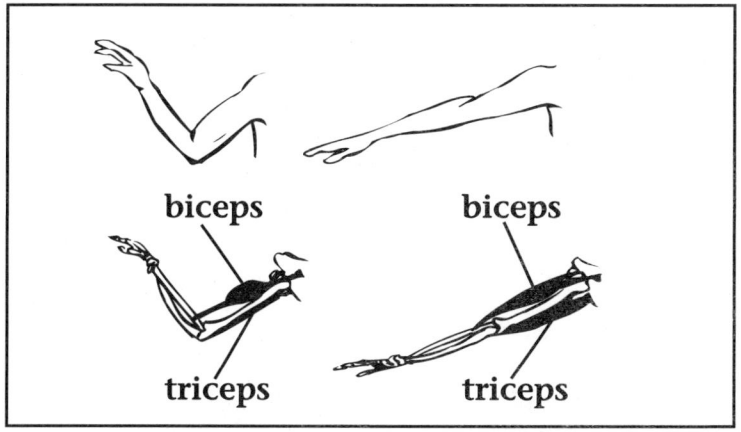

The **biceps** muscle in the arm contracts (shortens) to pull up the lower part of the arm. The **triceps** muscle contracts to pull it back and make the arm straight.

Doing exercises regularly can make muscles bigger and stronger, but a lot of care is needed to prevent **injury** during exercise. Muscles do not **fracture** (break) like bones, but they can tear like fabric. This is very painful and makes movement difficult.

Human muscles

1. What is the main job that **muscles** do for us?

   _____

2. In what way are muscles not like **bones**?

   _____

3. What joins muscles to bones? _____

4. How do muscles know what to do?

   _____

   Do muscles push or pull? _____

5. If a swimmer says he has a 'frozen shoulder', what does he mean?

   _____

6. Complete this passage.

   When a **muscle** is working, it _____, which means that its

   length gets _____. When a muscle is not working, it

   _____ and its length goes back to _____.

7. When a human arm is **relaxed**, is it straight or bent? _____

   Is a bent arm relaxed? Explain why. _____

   _____

8. Name the two muscles in the human arm.

   _____

Muscles and bones are different. Make a chart with words and pictures to show some ways that they are different (for example, bones break but muscles tear).

SCHOLASTIC DEVELOPING SCIENCE LANGUAGE for Living Things with 8–9 year olds

Human muscles

1. What is the main job that **muscles** do for us?

_____

2. **Bones** are hard. Are muscles hard or soft? _____

3. What joins muscles to bones? _____

4. Do our brains tell our muscles to push or to pull? _____

5. A swimmer might get a 'frozen shoulder'. What does this mean?

_____

6. When a muscle is working, does it become shorter or longer?

_____

7. Draw a circle around each word that describes a muscle that is not working.

      stretched   perplexed   relaxed   compressed

8. Which muscle in your upper arm lifts your hand to touch your face?

_____

Muscles and bones are different. Make a chart like this, with words and pictures, to show three ways that they are different (for example, bones break but muscles tear).

Muscles	Bones

Missing muscle words

Use these words to fill the gaps in the sentences below.

relaxed	pull	shorter	contracting	tendons	muscles

Bones cannot move on their own. They are made to move by

_____, which are attached to bones by _____.

Muscles do not push, they only _____. They do this by

_____, which means that they become _____.

When muscles are not pulling, their length returns to normal and

the muscle is _____.

Use these clues to find words that fill the spaces in the box below.

1. Our teeth are in this.
2. An arm joins the body at this joint.
3. There are feet at the ends of these.
4. This is the front part of the lower jaw.
5. These move when we blink.
6. We see with these.
7. There are hands at the ends of these.

1.								
2.								
3.								
4.								
5.								
6.								
7.								

If your answers are correct, you should find another science word.

Write it here. _____

SCHOLASTIC DEVELOPING SCIENCE LANGUAGE for Living Things with 8-9 year olds

Bend and stretch

Label this diagram of a human arm. Write captions in the boxes to explain how the muscles work to bend the arm at the elbow.

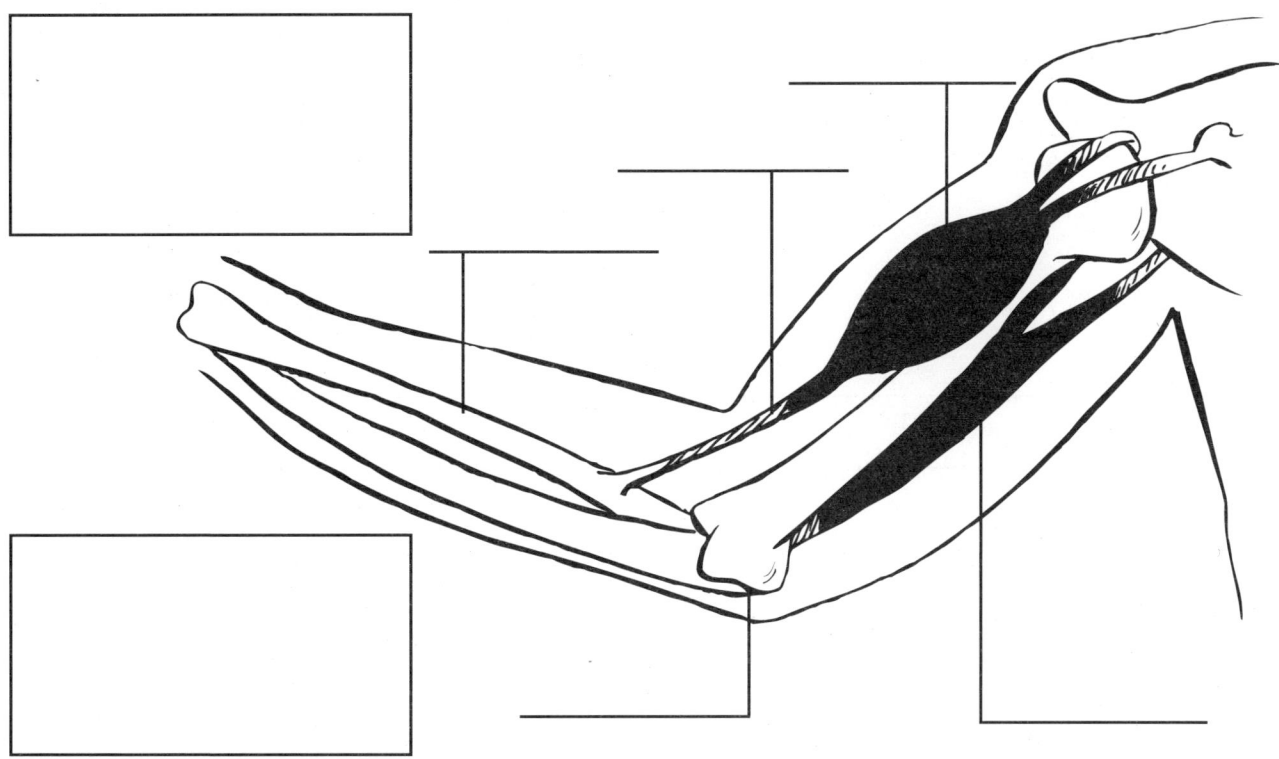

The prefixes of the two words **biceps** and **triceps** mean particular numbers. Think of things that you can ride on.

Which are the letters? ____ and _____.

What are the numbers? ____ and ____.

Find out what this tells us about the biceps and triceps muscles.

bone	triceps	tendon	biceps	elbow

COVER THESE INSTRUCTIONS WHEN PHOTOCOPYING.

Note for teacher
The words in the box above are the labels needed for the arm diagram. They can either be given to the children or be covered up when the page is photocopied, as you prefer.

Muscular activities

The muscles in our legs enable us to move forwards and backwards when we walk or run. To make this happen, pairs of muscles work together in a controlled way. Sometimes we move parts of our bodies without moving from where we are. All of this needs muscular activity.

Write words to go with these drawings.

Write a sentence about the muscles that help us to chew, saying where they are and what they are called.

_____

The human body contains a very special muscle that contracts and relaxes about 70 times every minute.

What is the name of this organ? _____

What is its main job? _____

_____

SCHOLASTIC DEVELOPING SCIENCE LANGUAGE for Living Things with 8–9 year olds

Habitats

Plants and animals live all over the Earth in a great **variety** of places. These places are called **habitats**. Some habitats are warm, while others are cold. Some are dry, while others are wet. These words (wet, dry, warm, cold) are called the **conditions** of each habitat. Because of their different conditions, different habitats are suitable for different living things. The science word for a living thing, whether it is an animal or a plant, is **organism**.

 **Sticklebacks** live among water weeds in ponds. **Woodlice** live in dark, damp places such as under rotting pieces of wood.

Habitats such as streams, woodland, rock pools and mountain peaks are **natural**. Some other habitats, such as an aquarium or a house, are not natural: they have been made by people.

Large habitats such as forests contain smaller habitats. A single tree is a habitat, and so is the carpet of leaves on the forest floor.

Other conditions of habitats can vary. One habitat may have lots of sunlight falling onto it, another may be dark. One habitat may give a lot of protection, another not much protection. A habitat may be changed by weather and the seasons, by the tides of the sea, or even by the organisms that live in it.

The organisms that live in a certain habitat need to have special **features** to help them **survive** in that habitat. Polar bears have thick coats of fur to keep them warm in their very cold habitat. Plants that live in very dry habitats have special leaves to keep as much water (and dissolved minerals) as possible.

Habitats

1. What is the scientific name for a place where plants and animals live? _____

2. Name two things that live in a pond.

    _____   _____

3. Explain what a **natural habitat** is.

    _____

    _____

4. Name one large habitat and one small habitat.

    _____

5. Name three **conditions** that may be different in different habitats.

    a) _____

    b) _____

    c) _____

6. Do you think a dolphin could survive in a dry place? Explain your answer. _____

    _____

7. Name two causes of change in a habitat.

    _____   _____

Write a letter to a friend to tell him or her about an organism that lives in a habitat near you. You should name the organism and say where its habitat is. Most important, tell your friend why you think the organism is well suited to this habitat.

Habitats

1. What is the scientific name for a place where plants and animals live? _____

2. Name two **organisms** that live in a pond.

_____

3. Are these natural habitats? Answer 'Yes' or 'No'.

A desert _____ An aquarium _____

Are these habitats made by people? Answer 'Yes' or 'No'.

Moorland _____ A quarry _____

4. Name one large habitat and one small habitat.

_____

5. Underline the conditions that may be different in different habitats.

temperature   amount of shade   wetness

6. You may have seen pictures of a 'beached whale'. What has happened, and how could people help the whale?

A 'beached whale' is _____

_____

People could help by _____

_____

7. Name two causes of change in a habitat.

_____   _____

Write a letter to a friend about an organism that lives near you. Tell your friend the name of the organism, where its habitat is and why you think that the organism is well suited to living there.

Changing places

Draw a line to connect each **organism** to its **habitat**. Remember that different organisms may live in the same habitat.

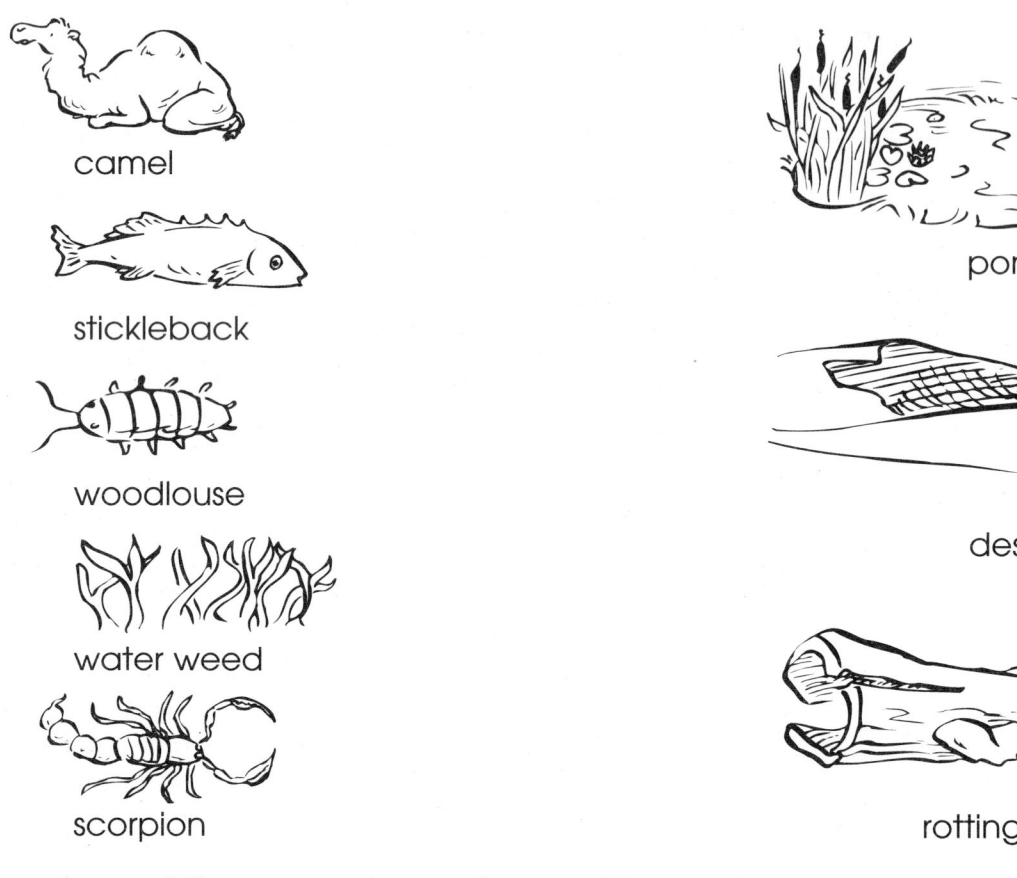

camel

stickleback

woodlouse

water weed

scorpion

pond

desert

rotting wood

Choose one of the organisms above. Sketch it in its habitat. Write a sentence to explain why its particular habitat is suitable for it.

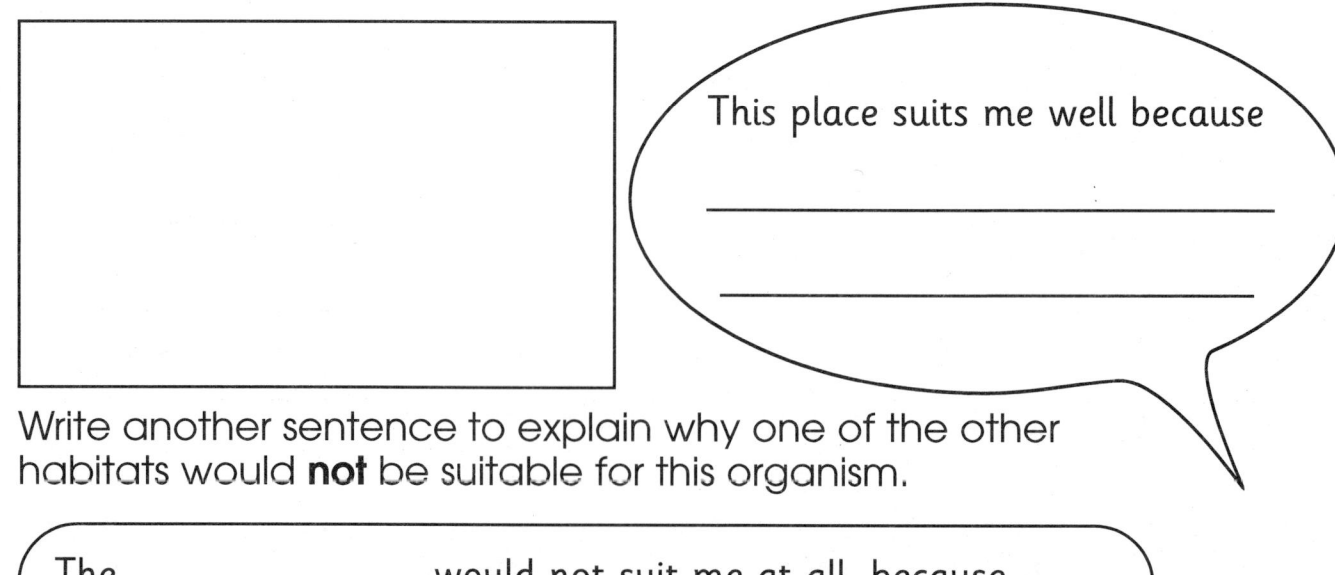

This place suits me well because

_____

_____

Write another sentence to explain why one of the other habitats would **not** be suitable for this organism.

The _____ would not suit me at all, because

_____

What lives where?

Choose the correct words from the box to complete the paragraph below.

change	habitat	aquarium	house	pets
woodland	zoo	protect	small	people

A place where organisms live is called a _____. Some of these places, such as _____, are natural. Others, such as a _____, have been made by people. Dead leaves on a forest floor make a habitat that is _____ in size. Habitats do not stay the same. They _____ as time passes.

You should have five words left in the box. Use these five words and other words to make three more sentences to tell your friends about organisms and habitats.

1. _____

_____

2. _____

_____

3. _____

_____

Scrambled words

Here are some words about **habitats**. The letters of each word are mixed up. Unscramble the letters and write the correct words.

g i l h t _____   r w e h e t a _____

m i a l s n a _____   r t a e w _____

s p a n t l _____   d n o c i i n s t o _____

n e g a c h _____   t h a r w m _____

r a l a n u t _____   e t w _____

Write three sentences to tell your friends about **habitats** and **organisms**. Use some of the words in the list above.

1. _____

_____

2. _____

_____

3. _____

_____

Think of three more words about habitats. Mix up the letters and write the scrambled words here. Ask your friends to try to unscramble them.

1. _____   2. _____   3. _____

COVER THESE INSTRUCTIONS WHEN PHOTOCOPYING.

Note for teacher
The final task on this page could be developed into a class game. The children could cut out the three scrambled words. You could collect all of the puzzle words and write them out on the board or flip chart, one by one, for the class to solve.

True or false?

Write 'True' or 'False' at the end of each sentence.

1. Plants move about to live in a variety of places. _____

2. Camels live in dark, damp places. _____

3. Temperature does not affect a habitat. _____

4. Habitats are not always the same. _____

5. Mountain-tops are made by people. _____

6. Organisms are musical instruments. _____

7. A stream is a natural habitat. _____

8. Fallen leaves on a forest floor are a natural habitat. _____

9. Several different organisms can live in the same habitat.

_____

Can you change each false sentence to make it true?
Write your new sentences here.

1. _____

2. _____

3. _____

4. _____

5. _____

Discuss these two statements with your friends or your teacher. Are they always true, always false, or sometimes true and sometimes false?

Forests are made by people. _____

Deserts are made by people. _____

Habitat puzzle

Use the clues to solve this habitat puzzle.

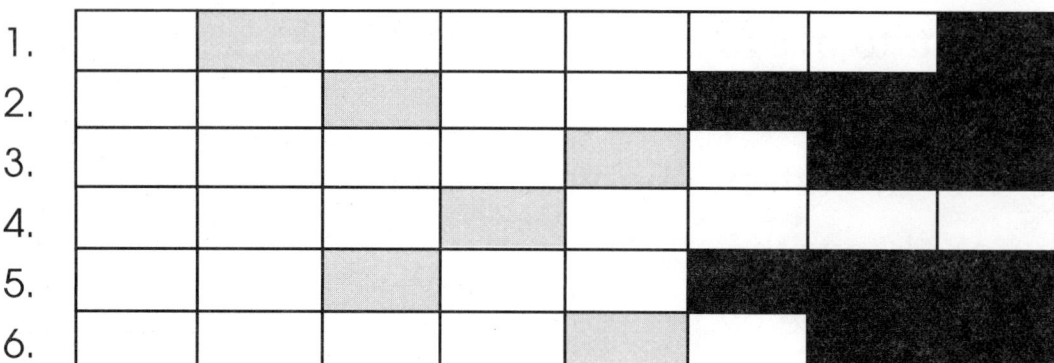

Clues

1. There are four of these in one year.   s_____

2. Plants and animals live on this planet. E_____

3. There are many trees in this.   f_____

4. These creatures live under rotting wood.   w_____

5. All living things need this liquid.   w_____

6. These grow on trees. l_____

Unscramble the letters in the shaded squares to find the name of a habitat that can be very hot or very cold.

Write the name of this habitat here. _____

Find out more about this habitat. Write a factfile about what lives there.

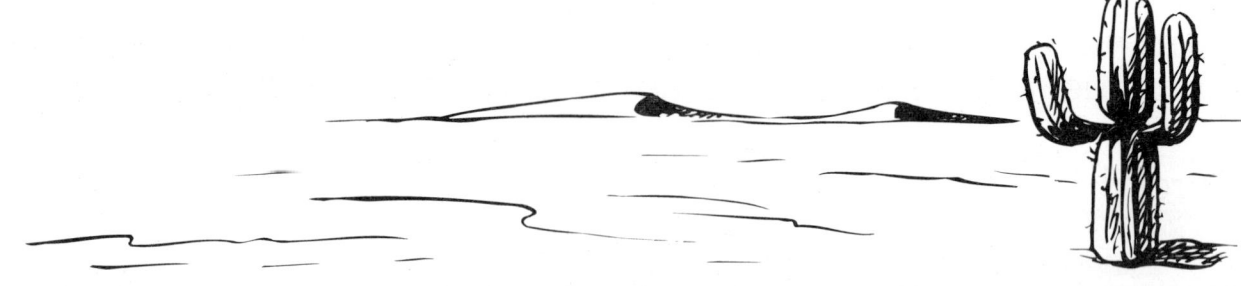

COVER THESE INSTRUCTIONS WHEN PHOTOCOPYING.

Note for teacher
Cover up these answers before photocopying the page. The answers to the puzzle are: 1. seasons 2. Earth 3. forest 4. woodlice 5. water 6. leaves. The word hidden in the shaded squares is 'desert'.

S C H O L A S T I C   DEVELOPING SCIENCE LANGUAGE for Living Things with 8–9 year olds

Keys

How do we **recognise** each other? When you see a friend coming towards you, how do you know that the person is your friend? One way to know is by looking at the person's **features**, such as hair or **height**, and matching them with what you remember about your friend. This happens very quickly in your brain. When we get closer, we can match the **colour** of the eyes, the **shape** of the nose, the **pattern** of teeth.

This person is tall, like my friend. It could be my friend.
This person has black hair, like my friend. It could be my friend.
This person has long hair. My friend has long hair. It could be my friend.
That is enough checking. This is my friend.

A **key** is the science word for a set of questions and answers that help us to recognise or **identify** a particular living thing, object or material. We use keys all the time in our lives, often without thinking. We go home to the same house every day and use the same bus or bike. If someone asked us 'Which one is your house?' or 'Which one is your bike?', we would have to describe some **features** of the house or bike. We use these features to **identify** our house or our bike by picking out the **differences** between it and the similar houses or bikes that are nearby.

In science, keys are often used to identify animals and plants. They are also used to identify things that are not living, such as rocks and metals.

Keys

1. What does a **key** mean in science? _____

_____

2. What does a **feature** mean in science? _____

_____

3. Describe three **features** of yourself.

a) My hair is _____

b) _____

c) _____

4. Cross out the incorrect word in each set of brackets.

A **key** helps us to **recognise / reorganise** someone.
A **key** helps us to **identify / liquify** something.
A **key** helps us to **sort / stick** similar things.
A **key** helps us to **separate / solidify** things.
A **key** helps us to **spot / lose** someone in a crowd.

5. One feature is not enough to **identify** a person. Explain why not.

_____

_____

6. Which three features would help us to identify a pencil?

_____

7. In science, what are keys mostly used for?

_____

Write a list of six features to describe a friend, pet or toy. Use whole sentences and try to give **detail**. It is more useful to write 'My friend has long, brown hair' than to write 'My friend has hair'.

SCHOLASTIC DEVELOPING SCIENCE LANGUAGE for Living Things with 8-9 year olds

Keys

1. In science, a **key** means a set of questions and answers that help us to do something. What does a key help us to do?

_____

2. Cross out the incorrect word in each sentence.

A **key** helps us to **recognise / reorganise** someone.
A **key** helps us to **identify / liquify** something.
A **key** helps us to **sort / stick** similar things.
A **key** helps us to **separate / solidify** things.
A **key** helps us to **spot / lose** someone in a crowd.

3. A **feature** in science usually means some part of the body of an animal or plant. If you are trying to identify a person, the colour of that person's shirt is not a good feature to use. Explain why not.

_____

4. Describe three features of yourself.

My hair is _____. My eyes are _____.

My _____

5. If you want to identify your goldfish, it is not always enough to say that it is gold. Why not?

_____

6. What three features would help us to identify a pencil?

_____ _____ _____

7. In science, keys are often used to identify animals and plants. True or false? _____

Write a list of six features to describe your friend, pet or toy. Write whole sentences and try to give **detail**. It is more useful to write 'My friend has **long, brown** hair' than to write 'My friend has hair'.

What shape?

Pretend that you do not know what shape this is:

Here is a key to help someone find out what the shape is called. Draw a ring around your answer to each question, and follow the arrows.

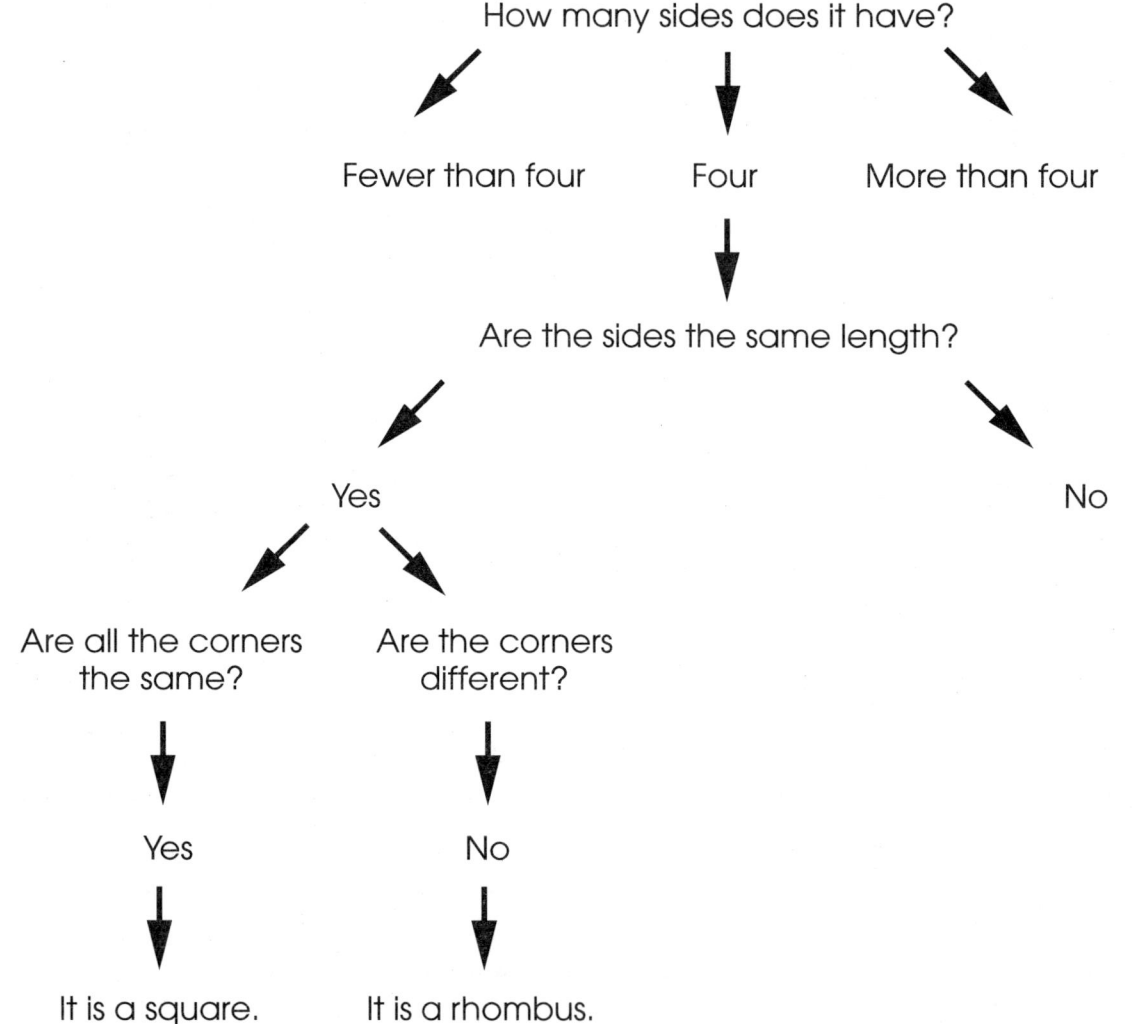

How many sides does it have?

Fewer than four   Four   More than four

Are the sides the same length?

Yes   No

Are all the corners the same?   Are the corners different?

Yes   No

It is a square.   It is a rhombus.

Make a key to help someone identify two other shapes, perhaps a triangle and a rectangle.

Note to teacher
This activity can be linked to work on the characteristics of shapes in mathematics lessons.

What is it?

Pretend that you do not know what these four vehicles are.

 A   B    C    D

Here is a key to help someone identify the four vehicles.
Draw a ring around your answer to each question.

Start by trying to identify vehicle A.

How many wheels does vehicle A have?

Answer: 1   2   3   4   more than 4

How many seats does vehicle A have?

Answer: 1   2   3   4   lots

Now use your answers and this table to find what vehicle A is called.

Number of wheels	Number of seats	Name of vehicle
2	1	Bicycle
3	1	Tricycle
2	2	Tandem
4	4	Car

Vehicle A is a _____

Use the key to identify the other vehicles.

Vehicle B is a _____

Vehicle C is a _____

Vehicle D is a _____

On another sheet of paper, make your own key to help someone identify different types of vehicle.

Making a key

Your friend has a pet called Ben. Think about how you could make a key to find out what type of pet Ben is.

Which of these would be a good question to ask first?

1. Does Ben have eyes?
2. Does Ben have a mouth?
3. Does Ben have legs?

Write a sentence to explain why you think each question is good or silly. The first one has been done for you.

Question 1   is a silly question because all pets have eyes, so it does not help.
_____

Question 2   is
_____

Question 3   is
_____

If the answer to Question 3 (about legs) is 'Yes', which of these questions would be a good one to ask next?

1. Does Ben have one leg?
2. Does Ben have two legs?
3. Does Ben have three legs?
4. Does Ben have four legs?

Explain whether you think these questions are good or silly.

Question 1 is **good / silly** because _____

Question 2 is **good / silly** because _____

Question 3 is **good / silly** because _____

Question 4 is **good / silly** because _____

If the answer is 'two legs', what
sort of pet animal might Ben be? _____

SCHOLASTIC DEVELOPING SCIENCE LANGUAGE for *Living Things* with 8-9 year olds

A key for plants

This is a different sort of key, but it still depends on asking good, sensible questions and following each answer to go to the next question.

These plants do not have flowers yet.
Can you identify them from their leaves?

A   B   C

Choose one answer each time and follow the instructions.

Question 1: How many leaves are on each stem?

If your answer is **3** go to Question 2.

If your answer is **1** go to Question 3.

Question 2: Are the edges of the leaves smooth or jagged?

If your answer is **smooth** it is the leaf of a clover.

If your answer is **jagged** it is the leaf of a buttercup.

Question 3: Does the edge of the leaf have sharp points?

If your answer is **yes** it may be the leaf of a holly.

If your answer is **no** it is the leaf of a daisy.

You should now be able to identify the three plants. Complete these sentences:

Plant A is a _____.

Plant B is a _____.

Plant C is a _____.

A key for birds

What birds are these?

A    B    C

Choose an answer each time and follow the instructions.

Question 1: Is the beak long or short?

If your answer is **long** go to Question 3.

If your answer is **short** go to Question 2.

Question 2: Is the beak straight or like a hook?

If your answer is **straight** it might be a warbler.

If your answer is **like a hook** it is an eagle.

Question 3: Is the beak straight or with a small bend?

If your answer is **straight** it is a heron.

If your answer is **small bend** it is a curlew.

Use this key to identify the three birds.

Bird A is a _____.

Bird B is a _____.

Bird C is a _____.

SCHOLASTIC DEVELOPING SCIENCE LANGUAGE for Living Things with 8–9 year olds

Food chains

All living things need food. Food gives them the materials and energy they need to build their bodies. Plants are called **producers** because they make food using sunlight and the air around them. Animals are called **consumers** because they either eat plants or eat other animals that have eaten plants.

The links between living things and other living things that they feed on are called **food chains**. Here is an example. **Energy** from sunlight enables grass to  grow. Cows eat the grass, which enables them to grow. Humans drink milk from cows and eat their meat (beef). In this way, humans get some of the energy that comes from the Sun.

We can show this food chain by drawing a diagram. The arrows show how **energy** moves along the chain.

grass (producer)   cow (first consumer)   human (second consumer)

This is a short food chain. Some other food chains are longer, with more consumers. For example:

seaweed   limpet   starfish   seagull

Seaweed is the producer. The starfish and seagull are **predators**. They **prey on** (kill and eat) other animals that are before them in the food chain. The seagull preys on the starfish, which preys on the limpet. The limpet is a **prey** to the starfish, which is a prey to the seagull. The starfish is both a predator and a prey.

Food chains

1. Why do living things need food? _____

_____

Which living things produce food? _____

What do **producers** need to make food? _____

2. In the seagull food chain, name the producer and the first consumer.

The producer is _____. The first consumer is the _____.

3. What kind of living thing always starts a food chain?

_____

4. Which of these words best finishes the sentence below?

    consumers   predators   producers

Green plants are _____

5. All the plants and animals in a food chain are **visible**. What is the **invisible** factor that flows through the chain? _____

6. These four living things are linked by a food chain.

lettuce    snake    slug    frog

Frogs eat slugs. Snakes eat frogs. Slugs eat lettuces.
Write down the food chain that connects the four organisms.

_____ ➔ _____ ➔ _____ ➔ _____

Draw two food chains:
(a) a food chain connecting you and potatoes
(b) a food chain connecting a tiger, grass and a gazelle.

Food chains

1. Why do living things need food?

_____

Which living things produce food? _____

What do **producers** need to make food? _____

2. Cross out the incorrect words in each sentence.

(a) A producer **makes food / eats food**.
(b) A consumer **makes food / eats food**.

In the seagull food chain,
which is the producer? _____

3. Is this sentence true or false?
Every food chain starts with a consumer. _____

4. Tick the best word to finish this sentence.

Green plants are   consumers ☐   predators ☐   producers ☐

5. Something **invisible** (you cannot see it) flows along a food chain
from start to finish.

Is it **energy** or **air** or **oxygen**? Draw a circle around the correct word.

6. These drawings show four living things.

lettuce   snake   slug   frog

Fill in the gaps in this food chain.

lettuce   ➤ _____   ➤   frog   ➤ _____

Draw two food chains:
(a) a food chain connecting you and wheat (used to make bread)
(b) a food chain connecting grass, a tiger and a gazelle.

Pond life

Draw a line to match each word on the left to its correct definition on the right.

Consumers   animals that eat other animals

Producers   animals that eat plants or animals

Predators   animals that are eaten by other animals

Prey   green plants that make food

This food chain connects some living things in a pond.

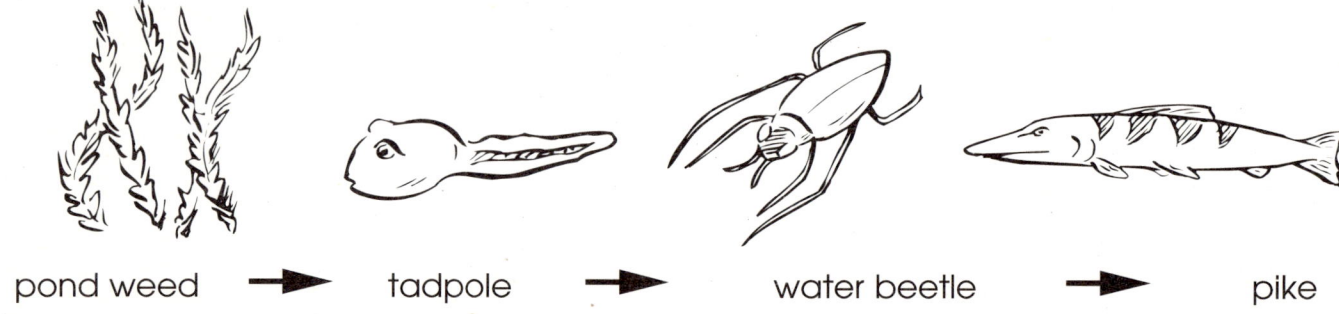

pond weed ➜ tadpole ➜ water beetle ➜ pike

Draw one line to match the name of each living thing to the best description of its place in the food chain.

pond weed   first consumer

tadpole   producer

water beetle   predator

pike   prey

Write a short account of the life of a tadpole as it grows into a frog.

Producers and consumers

grass robin cat seaweed

cod horse human daisy

dog seagull lettuce snail

cabbage tortoise willow greenfly

Write the name of each living thing in the correct box. One has been written for you.

Producers	Consumers
grass	

Some consumers do not directly eat producers, but they still depend on the energy that producers have turned into food.

Is this statement true or false? _____

Explain your answer. Write about an actual food chain if that helps.

_____

_____

_____

Ordering a food chain

Arrange each group of living things into a food chain. Always start with a producer.
Draw a red circle around each first consumer.
Draw a blue circle around each second consumer.

| man |  grass |  cow |

The food chain for these living things is:

| | → | | → | |

| grass |  fox |  rabbit |

The food chain for these living things is:

| | → | | → | |

| cabbage |  caterpillar |  blackbird |

The food chain for these living things is:

| | → | | → | |

Ordering a food chain

2

Arrange each group of living things into a food chain or **food web**.
Always start with a producer.
Draw a red circle around each first consumer.
Draw a blue circle around each second consumer.

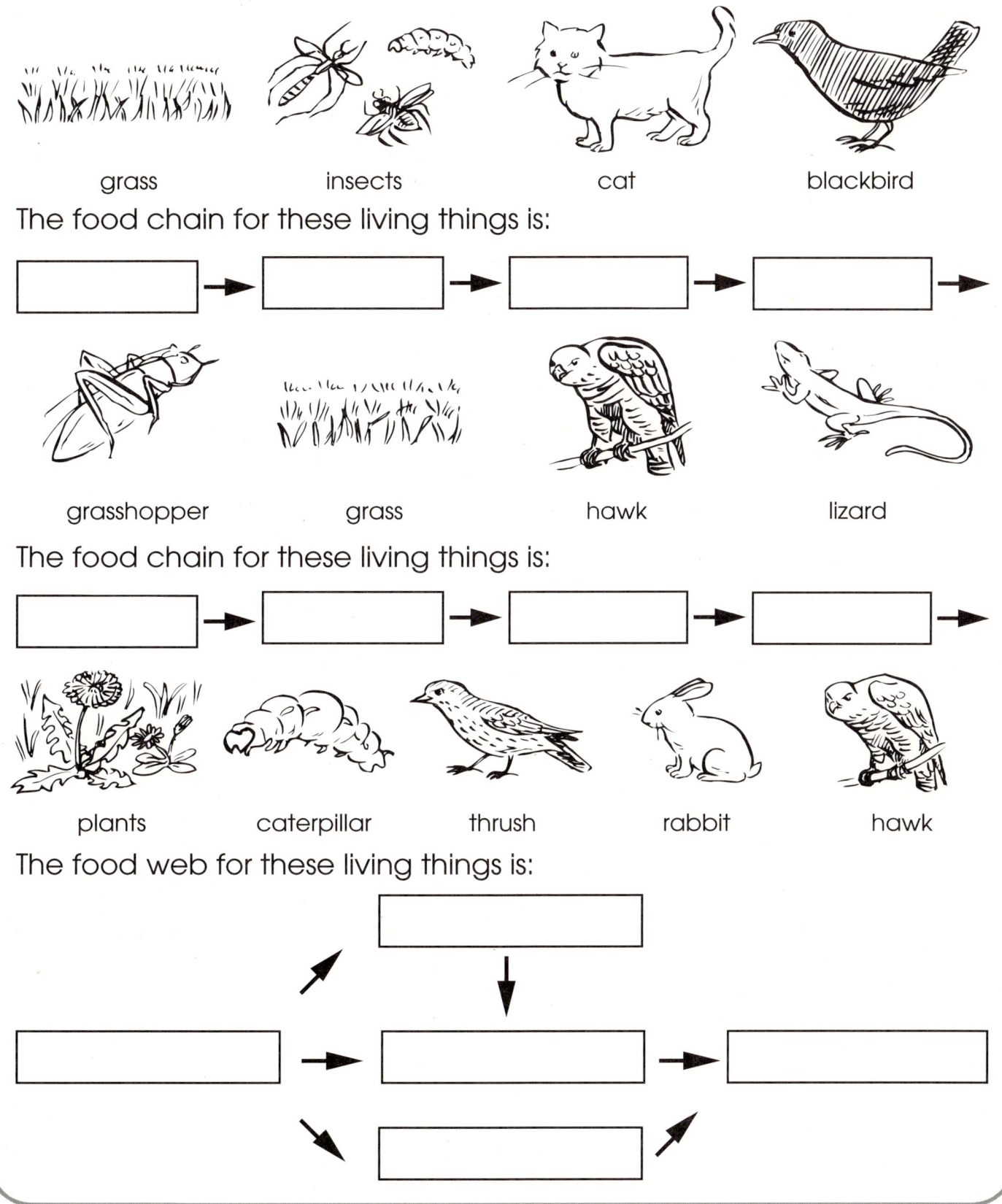

grass   insects   cat   blackbird

The food chain for these living things is:

grasshopper   grass   hawk   lizard

The food chain for these living things is:

plants   caterpillar   thrush   rabbit   hawk

The food web for these living things is:

Food chains crossword

Can you solve this crossword?
Most of the answers have to do with food chains.

Clues across

1. A watery environment.
4. Opposite of dry.
5. This comes from the Sun and helps us to see.
6. A mammal's coat.
8. A wide-eyed night bird.
10. An animal that is eaten by another animal.
11. A good worker in the soil.

Clues down

1. An animal that eats other animals.
2. This swims on the answer to 1 across.
3. A series of links.
6. Birds do this.
7. This anchors a plant.
9. Some food chains connected together make a food _ _ _.
10. Two letters meaning 'afternoon'.

Draw a food chain that involves the living things named in 8 across and 11 across. You will need to think of two other living things to complete the chain.

Name the producer in this food chain. _____

Name one predator in this food chain. _____

SCHOLASTIC DEVELOPING SCIENCE LANGUAGE for Living Things with 8-9 year olds

Conservation

A very long time ago, there was nothing living on our planet
Earth. There were no trees, no flowers, no insects, no reptiles, no
fish, no birds, no humans.

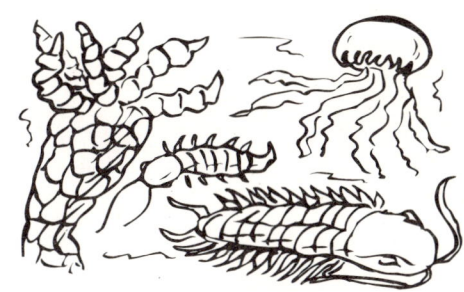

Slowly, over many, many years, some
living things began to appear in the
water – and then on the land.
Gradually, the **cycle of life** was
established: birth, life, death, birth.

However, the **conditions** on Earth did not stay the same.
Habitats changed and some groups (**species**) of living things
were not able to **adapt** to the changes, so they died out. They
became **extinct**, never to live again.

Some changes in habitats have occurred
naturally, because of **climate**. Other changes
have occurred because of humans. As people
have wanted more places to live (towns and
cities) and wanted more materials (coal, oil and
wood) from the Earth, they have often destroyed
the habitats of other living things. Some species
have become **rare**: they have very few
individuals left to carry on their life cycle. Other
species have become extinct.

Today, many people understand that it is not good for humans to
behave so selfishly that they cause other species to become rare or
extinct. These people are involved in **conservation**, which means
trying to protect living things from harm so that they can live their
own cycle of life. We can all help to protect life on Earth.

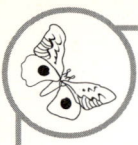

Conservation

1. What is on the planet Earth now that was not there a very

long time ago? _____

2. After many years, what happened on the Earth?

_____

_____

3. What are the four stages in the cycle of life?

1. _____ 2. _____ 3. _____ 4. _____

4. What is the meaning of **rare species**?

_____

5. What is the meaning of **extinct species**?

_____

6. What kind of changes in habitat might have caused a species to become rare or extinct?

_____

_____

Use books or CD-ROMs to find out about a species that is rare or extinct. Write a factfile about it.

Conservation

1. What things are on the planet Earth now that were not there

a very long time ago? _____

2. After many years, something began to happen. What was this?

_____

3. What are the four stages in the cycle of life?

1. _____ 2. _____ 3. _____ 4. _____

4. How many individuals are there in a **rare species**? Tick one answer.

many ☐   quite a lot ☐   very few ☐   none ☐

5. What does **extinct species** mean?

_____

6. Some species became extinct a long time ago. Why do you think
this happened? Draw rings around the likely reasons in this list.

            no shops   no schools   not enough food

        too many predators   no place to shelter

Write a factfile about one species that is rare or extinct. Use books or
CD-ROMs to help you.

Careless humans

Of all living things, we humans are the worst offenders in destroying the habitats of other living things.

Describe three ways in which we destroy habitats.

_____

_____

_____

Think about where you live. Write about two areas in your environment where human activity has meant that conditions are not good for living things. For example, you might write about a pond, a field or a street.

_____

_____

_____

How do you think local people could change these two areas so as to improve the conditions for living things? Describe what people could do to **conserve** life in each area.

_____

_____

_____

Write a letter to someone important to say why everyone should try to help with conservation. Use what you have written above to help you write your letter:
- Start by explaining how we destroy habitats.
- Describe two local areas where habitats have been destroyed.
- Explain how these areas could be changed to help conserve life.

When do birds need food?

This chart tells us about ten birds. It shows in which months they need a high (H) amount of food, a medium (M) amount or a low (L) amount.

	Jan	Feb	Mar	Apr	May	Jun	Jul	Aug	Sep	Oct	Nov	Dec
Blackbird	H	H	H	M	M	M	M	M	L	L	M	H
Blue tit	H	H	H	H	M	M	H	H	L	L	M	H
Chaffinch	H	H	H	H	M	M	H	H	L	L	L	M
Collared dove	H	H	H	H	H	H	H	M	L	L	L	L
Dummock	H	H	H	H	H	M	L	L	M	H	H	H
House sparrow	H	H	H	H	H	H	M	L	L	L	M	H
Robin	H	H	H	M	M	M	M	L	L	M	M	M
Song thrush	H	H	H	M	L	L	L	L	L	L	H	H
Starling	H	H	M	L	L	M	M	M	M	L	L	H
Wren	H	H	H	M	L	L	L	L	L	L	L	L
H				5								
M				4								
L				1								

In the three rows at the bottom of the chart, write numbers to show how many birds need each level of feeding (H, M or L). The numbers for April are shown. Make sure you understand how these numbers for April were calculated.

Here is another chart to show the information in a different way. The data for April is shown for you. Fill in the information for the other months.

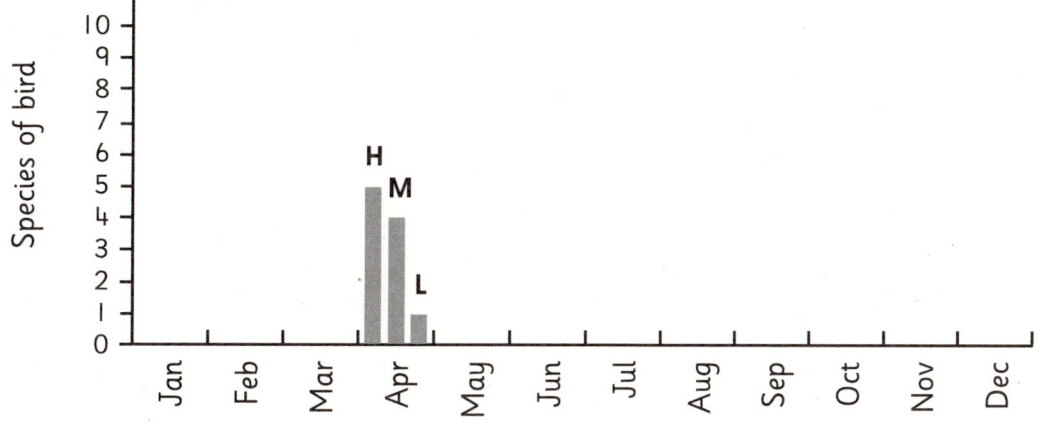

Which of the two charts do you find more useful to help you understand the feeding needs of birds?

A nest box for a robin

Here is part of a plan to make a nest box for a robin.
Complete the plan so that someone else can follow it.

You will need:

1. One strip of balsa wood. ● How long? _____   ● How wide? _____

2. Strong glue   **3.** Pencil   **4.** Ruler   **5.** Hacksaw   **6.** Hammer and two nails

What to do

Use a pencil and a ruler to mark off the different parts of the box.

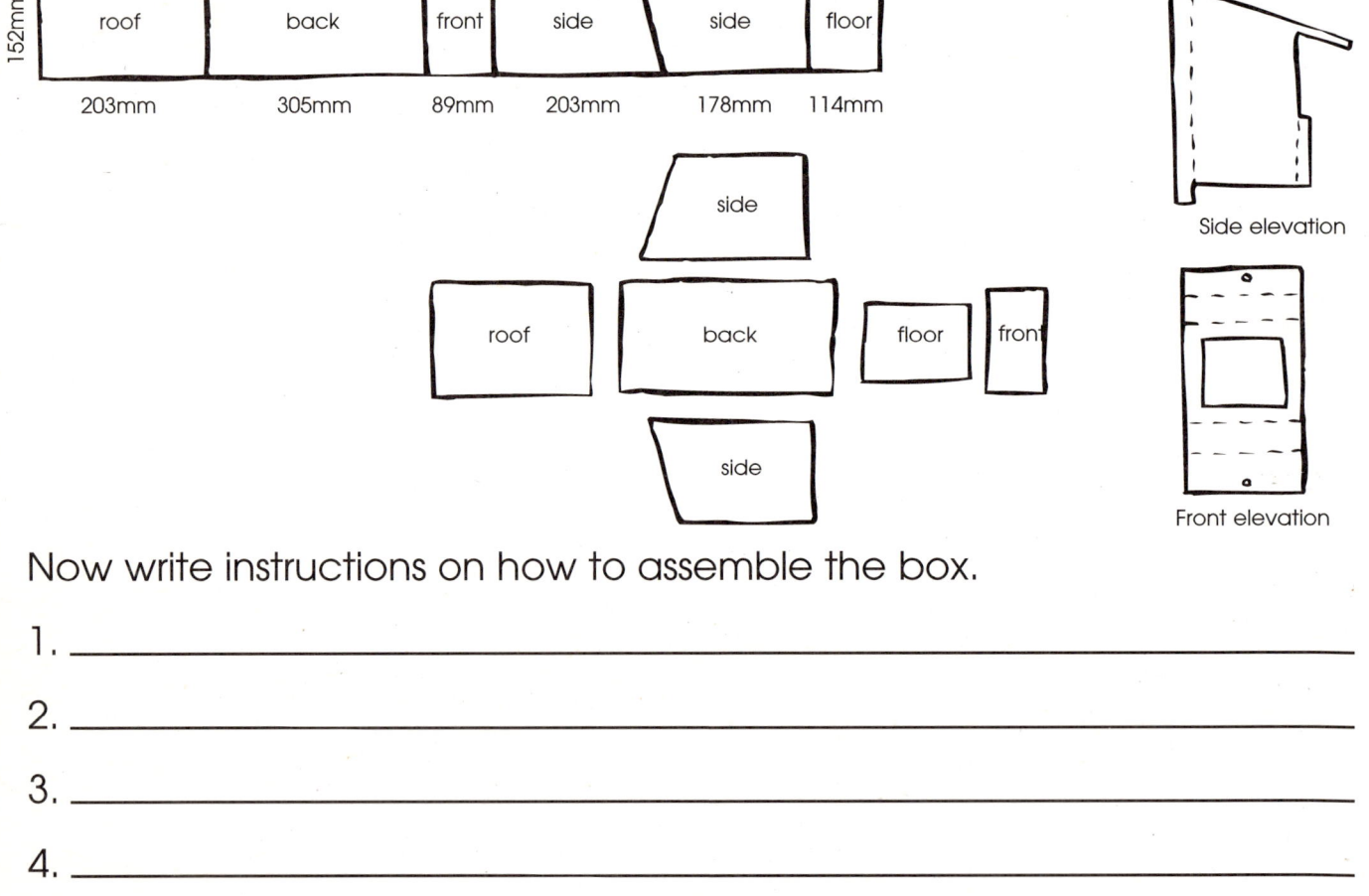

Now write instructions on how to assemble the box.

1. _____

2. _____

3. _____

4. _____

5. _____

Where will you hang the box to make the robin feel safe?

_____

Will you paint the box inside and outside? Explain your answer.

_____